THE FUTURE OF
CONSERVATION IN AMERICA

THE FUTURE OF
CONSERVATION
in America
A CHART FOR ROUGH WATER

GARY E. MACHLIS

JONATHAN B. JARVIS

WITH A FOREWORD BY TERRY TEMPEST WILLIAMS

THE UNIVERSITY OF CHICAGO PRESS

CHICAGO AND LONDON

The University of Chicago Press, Chicago 60637
The University of Chicago Press, Ltd., London
© 2018 by Gary E. Machlis and Jonathan B. Jarvis
Published 2018
Printed in the United States of America

27 26 25 24 23 22 21 20 19 18   1 2 3 4 5

ISBN-13: 978-0-226-54186-0 (cloth)
ISBN-13: 978-0-226-54205-8 (paper)
ISBN-13: 978-0-226-54219-5 (e-book)
DOI: https://doi.org/10.7208/chicago/9780226542195.001.0001

Frontispiece: Yosemite Valley, Yosemite National Park. Photo credit:
Rob van Esch/shutterstock.com.

Library of Congress Cataloging-in-Publication Data
Names: Machlis, Gary E., author. | Jarvis, Jonathan B., author. | Williams, Terry
    Tempest, writer of foreword.
Title: The future of conservation in America : a chart for rough water / Gary E.
    Machlis and Jonathan B. Jarvis ; With a foreword by Terry Tempest Williams.
Description: Chicago ; Illinois : The University of Chicago Press, 2018. |
    Includes bibliographical references and index.
Identifiers: LCCN 2017058460 | ISBN 9780226541860 (cloth : alk. paper) |
    ISBN 9780226542058 (pbk. : alk. paper) | ISBN 9780226542195 (e-book)
Subjects: LCSH: Conservation of natural resources—United States. | Conserva-
    tion of natural resources—Political aspects—United States.
Classification: LCC S930 .M13 2018 | DDC 333.70973—dc23 LC record available at
    https://lccn.loc.gov/2017058460

♾ This paper meets the requirements of ANSI/NISO Z39.48–1992 (Permanence
of Paper).

CONTENTS

TOTALITY: A FOREWORD

TERRY TEMPEST WILLIAMS

On Monday, August 21, 2017, a solar eclipse crossed the United States of America, moving from Oregon to South Carolina, creating a swath of awe and wonder for the millions of Americans with their eyes focused upward. For those precious two and a half minutes of totality, it was a moment of grace.

My family and I watched the eclipse in Grand Teton National Park. From beginning to end, we were spellbound. We noted that when the moon's shadow took its first bite into the sun, sandhill cranes began trumpeting, their ancient sonorous cries marking the moment as a pair flew over us, close enough to touch. We paid attention to the quality of light, the movement of insects, the songs of birds. It was a gift simply to be present together, with the Teton Range at our backs.

A few minutes before totality, the temperature dropped dramatically, perhaps twenty degrees, the light deepened, and suddenly, without warning, we were encircled by a ring of sunsets. The world became very still. Twilight. Totality! Communal gasps. The black disc of the moon appeared as a dilated pupil focused on Earth. The sun's bright white corona against an indigo sky began spiking wildly, magically, with each second sparking a celestial phenomenon we had been imagining for months, years—for some, a lifetime. We were laughing and crying, at once, all of us standing eyes pointed upward. How to take it all in—I turned west for a glimpse of the shadowed Tetons and, then, quickly resumed my naked gaze to the total eclipse—and then, it was gone. The diamond of perfect light cut through the darkness, we put our protective glasses back on. It was over. I wanted more.

We continued to watch the movement of the moon reverse itself until the sun was restored to its full glory and the sandhill cranes marked that moment, too. On that day, I felt a unity of purpose and place I had almost forgotten in these turbulent times.

We live in a miraculous world.

Gary Machlis and Jon Jarvis remind us of this fact. *The Future of Conservation in America* is a clarion call for citizen engagement, an assurance that the open space of democracy remains open. Their chosen subtitle, *A Chart for Rough Water*, signposts their direct response to the threats we are facing. What they offer us is a visionary pragmatism that is clear, concise, and prescriptive. Their assumption is blunt: we all share a responsibility to honor and uphold the beauty of the American landscape, both human and wild. It is our duty, an essential form of patriotism. We could not have better guides. Between the two of them, Machlis, a respected scientist, and Jarvis, a former director of the National Park Service, they represent decades of service to conservation. They could not be more dedicated to or serious about finding a way forward, given where we find ourselves now in the midst of climate change, species extinction, and economic disparity with an administration in power who denies, deceives, and undermines ecological integrity and social justice—rough waters, indeed.

It is easy to fall into a sea of despair and risk drowning in our own despondency. But Machlis and Jarvis have no patience with this. The authors of *The Future of Conservation in America* believe in the core concept of "strategic intention" where "conservation actions (protecting a watershed, reforming a management policy, creating an educational program, organizing an event, and so forth) are most effective when they build the foundation and momentum for future action." They envision for the future a broader, younger, and more diverse constituency for conservation, one that sees the health of the land as our own. They see not just one story aligned with our national parks and monuments, largely white and privileged, but many stories embedded in these public places that illuminate our noble and wide histories: Harriet Tubman and the Underground Railroad; Stonewall Inn, the site of the LGBTQ community's protests and victories; César E. Chávez's sanctuary at La Paz in Keene, California, the center of their community organization, which lifted up the United Farm Workers. These touchstones of American democracy and direct action are now national monuments, part of our public commons that hold our varied natural and cultural histories where we can reflect on the character of this country we call home in the United States of America.

Bears Ears National Monument in Utah, established on December 28, 2016, is another example of what "conservation for the future" can become in the twenty-first century. The leadership of the Bears Ears Intertribal Coalition, composed of Hopi, Navajo, Ute, Mountain Ute, and Zuni Nations, shows us another model for understanding not only a sense of place but also an ethic of place. The Tribes advocated for the merging of traditional knowledge with Western science, calling for cooperative management among federal agencies for the first time in history. They were heard by President Barack Obama and government officials, Sally Jewell (former U.S. secretary of the interior) and Jon Jarvis, among them. It has been the beginning of a powerful rapprochement between Indians and the United States government.

But now, under the Trump administration, this very monument (alongside several others under review) are threatened, calling into question the Antiquities Act of 1906. The 1.3 million acres now protected, the site of hundreds of thousands of pre-Puebloan artifacts, may be sacrificed to oil and gas development in Utah's fragile red rock desert. This battle to reduce or rescind Bears National Monument will be fought in the federal courts. It is also being fought in the court of public opinion.

Not long ago, I was with Navajo elders, native leaders and medicine people in Bluff, Utah, members of Utah Diné Bikéyah, a nonprofit organization that has taken a significant leadership role in advocating for Bears Ears. I asked Willie Grayeyes, the chairman, who lives at Navajo Mountain, what to do with my anger in the face of all we stand to lose. He looked at me. His words were direct. "This is not a time for anger; it is a time for healing."

Our public lands and waters—deserts, forests, prairies, our national parks and monuments, wildlife refuges and free-flowing rivers, lakes, wetlands, and oceans—are our common ground, our natural inheritance to be passed on from one generation to the next. They are our soul-geographies, the landscapes of our imaginations, the seedbed of an ecological state of mind. We are not only inspired but healed by nature's sense of integrity, harmony, and wholeness. Each time I stand at the Needles Overlook in Canyonlands National Park in the midst of this vast erosional landscape carved and created through wind and water and time, deep time, I have this sensation of being very, very small, and yet very, very large, at once. The Navajo have a word for this kind of balance and beauty: *hózhó*. We are one with the universe.

Without a spiritual dimension to our work as conservationists, we are only working for ourselves, not the future, and certainly, not for future generations of all species.

Without a deeper understanding of the interconnectedness of life and the world we live in as people in place; and without recognizing and acknowledging how climate change and economic and racial inequalities have contributed to the societal traumas of Hurricanes Katrina, Sandy, and now Harvey flooding Houston, we are in more than rough waters—we are adrift in a hell of our own making.

"The national narrative is always evolving, and its arc must bend toward a fuller truth," these authors tell us. Conservationists today are truth-tellers: scientists, clergy, teachers, artists, writers, poets, doctors, nurses, healthcare workers, government workers, attorneys, judges, entrepreneurs, businessmen and -women, industrialists, educators, computer analysts, farmers, gardeners, contractors, counselors, union workers, laborers, activists, parents, and children. We are all conservationists if we care about justice and the quality of our lives within our communities.

Let this moment be an awakening from the dark chambers of denial, self-interest, and market-driven decision making. Let this moment be a time of recognizing how we belong to a great and abiding beauty that came before us and will survive us long after we are gone. Let this moment be a reckoning in which we realize there is no hope without action. And let our greatest actions be rooted in compassion for all species, not just our own.

This book in hand, *The Future of Conservation in America*, is more than a look at where we have been and where we are going environmentally. It is a personal reflection based on science and first-hand experience, a primer on how we can engage and be present in the service of life with all its complexities, how natural histories embedded in America's diverse landscapes can illuminate our diverse cultural histories. In this process of discovery, an atmosphere of greater empathy for one another is nurtured and cultivated. Perhaps, this is what Willie Grayeyes meant when he said, "This is not a time for anger; this is a time for healing"—an idea worth repeating in the name of understanding and ceremony.

The Great American Eclipse of 2017 felt like ceremony as it offered us a moment of unity with all eyes focused on something larger than ourselves. In that unity, even as I stood next to my father, now eighty-four years old, a man invested in a very different generational and political outlook, I felt what brought us together. Awe. Reverence.

Humility. We are no more and no less than the relationships that sustain us. Our joy of sharing this moment together in Jackson Hole, Wyoming, the landscape where our family has convened for decades, deepened our connection. Call it by its name: love. And our love for the larger community, in this particular place consisting of sage, aspens, lodgepole pines, sandhill cranes, grizzly bears, and human beings with the Teton Range as our spine, expands our spirits and humbles our hearts. The natural world is where we can locate our wider kinship. These are the bonds that tie us to eternity. They are also the bonds that when severed, leave us adrift in isolation and loneliness. Without the landscapes that have raised us, shaped us, informed who we are, we become refugees to a world we no longer recognize.

Machlis and Jarvis say, "A new and unified vision of conservation is required."

In this era of climate change, we are being given an unprecedented opportunity to come together around the world and embrace a transformational moment of courage, collective discipline, and devotion to survival on this beautiful, broken world. It is daunting. It is sobering. And it is essential. It can also be a time of immense creativity. What is at stake? Everything we hold dear from place to people to planet. No matter how hard one particular president and his administration is working against these ethical tides of care and compassion, the tide is turning. We must prepare and manage for change.

This is a stellar moment. This is our calling. This will be our healing, and this book can serve as a guide. For in the future of conservation is the future of our humanity.

THE FUTURE OF
CONSERVATION IN AMERICA

CHAPTER ONE

Watershed

The year 2016 witnessed two events significant to the future of conservation in America. One was the centennial of the National Park Service. The centennial was celebratory, marking the hundredth anniversary of its creation as the federal agency assigned to care for the growing number of America's national parks and historic sites.

The anniversary generated wide public interest and support for national parks, public lands, and conservation. It led to over 330 million visits to the National Park System, a record, and more than major league baseball, football, basketball, NASCAR, and the Disney amusement parks *combined*. President Obama established nine new national monuments during the centennial year, including the historic home of the National Women's Party in Washington, DC; the eighty-seven thousand–acre Katahdin Woods and Waters National Monument in Maine; and Bears Ears National Monument in southeastern Utah.[1] Other federal agencies, cities, state park systems, corporations, and many conservation groups joined the celebration.

Beyond the centennial activities, 2016 saw the federal government and many states advancing progressive and scientifically grounded agendas to confront the challenges of climate change. World conservation leaders welcomed a renewed American presence and leadership within the international conservation movement. Over two hundred countries signed the Paris Accord on climate change, including the United States. Conservation was gaining ground on many fronts.

The other momentous event was a national election of a new president and Congress. The election was divisive, profane, and intensely fought over the course of the year. Political expression was coarsened, ugly, and often malevolent. Donald Trump's rallies unleashed long-contained public anger, as he encouraged chants of "lock her up!" aimed at his opponent. Hillary Clinton cast Trump's supporters as "deplorables," and the far right and alternative fringe media amplified the viciousness of the campaign.

Attacks on conservation were a constant feature of the Republican side, framing conservation as "job-killing" or "government overreach." Specific targets included Environmental Protection Agency regulations, the Paris Accord, and the Obama administration's denial of permits for the Keystone Pipeline. Environmental and climate scientists were labeled as fraudulent (Trump had declared them "hoaxsters" as early as 2012).[2] President Obama's proclamations establishing national monuments were portrayed as unjust and ill-done takings of local rights. Members of Congress and local, vocal activists fervently disputed federal management of public lands. Opposition turned violent when armed militiamen and ranchers staged a hostile takeover of the Malheur National Wildlife Refuge in eastern Oregon, established in 1908 by President Theodore Roosevelt to protect habitat for birds.

These two events, the centennial and the election, converged on October 22, 2016, when candidate Trump visited Gettysburg National Military Park. The park commemorates one of the most devastating battles of the American Civil War, with fifty thousand casualties during three days of brutal fighting. It also commemorates a revered speech in American oratory, President Abraham Lincoln's Gettysburg Address. Historian David Voelker describes the scene of Lincoln's remarks:

Four and a half months later, the process of reburying the thousands of bodies that had been shallowly interred on the battlefield had begun but was not yet complete. In this sobering setting, Lincoln delivered a brief address to an audience of about 15,000 people, who interrupted him five times to applaud. Newspapers across the North also responded very favorably. Lincoln's comments that day, however, comprised only a brief moment in the cemetery's dedication. Prior to Lincoln's three-minute speech came music, a prayer, and a featured oration, a two-hour discourse delivered by Edward Everett, retired Massachusetts politician and former president of Harvard. While Everett's speech dwelled on the details of the battle, Lincoln attempted to give meaning to the events at Gettysburg, indeed to the Civil War itself, by speaking about the ideals for which he believed the Union stood.[3]

Using the same historic battlefield as stage and lectern, Donald Trump gave a speech that quickly "curdled into bitter resentment," reflecting a strident contrast to the sacrifice symbolized by the park and the ideals of Lincoln. Trump railed against his perceived enemies, particularly the media, the government, the opposition party, and the intellectual elite.[4] The irony of the setting was perhaps not noticed by the candidate.

The election was held less than three weeks later. A stunning amount of money had been spent—over $2.2 *billion* in the presidential race alone.[5] Nevertheless, overall voter turnout was modest, representing 58 percent of eligible voters. Donald Trump was elected president, winning the Electoral College while losing the popular vote by three million votes.

*
**

The clash of these two events—the centennial of the National Park Service and the election of Donald Trump as president of the United States—reflect much larger forces and divisions operating within American society. Conservation of natural resources such as forestlands and preservation of scenery such as mountains, hot springs, and waterfalls have long been traditions in America. The well-told story describes a movement initiated by 1890s' business interests (some linked to the railroads and firearms manufacturers) and wealthy individuals. The movement grew to include both of the major political parties and middle- and working-class Americans (mostly white) and has recently expanded to include persons of color, Native Americans, underrepresented communities, and the millennial generation.[6] Still, many Americans perceive conservation as driven by and for the economic and cultural elite.

There has always been an anti-elite and populist strain

in American politics, beginning with Andrew Jackson's rise, the contested election of 1824 (which he labeled "corrupt"), and Jackson's election as president in 1828. Current populism has recycled themes of previous populist risings—the People's Party of the 1890s, Huey Long's popularity in the 1930s, the racially divisive presidential campaign of George Wallace with his slogan "Stand Up for America" (1968), and Pat Buchannan's two presidential campaigns (1992 and 1996) with his claims that "the news media lies!" The political fuel driving these cycles of American populism has been working-class resentment at being disrespected, forgotten, and placed at unfair disadvantage in securing economic opportunity. In 2016, the resentment was loosely targeted at the intellectual and financial elites, scientists, the media, minorities and immigrants, and the Washington, DC–centered federal government.[7] The conservation movement (other than hunting and fishing groups) was largely included in this collection of resented interests.

The centennial and election in 2016 were both watershed moments in the nation's civic life. But events are not trends. The deeper patterns represented by the centennial and election matter far more than one anniversary and one voting cycle. Our focus is on the immediate challenge *and* the long view. The election of Donald Trump was a signal event; the deeper challenges of populism, anti-science attitudes, and resentment against government are a more profound and long-term concern. The long-simmering tension and now open conflict between those who view conservation as a shared, bipartisan, and vital national and global interest and the opposing rise of a populist, nationalist politics embracing class resentment, strident partisanship, and narrow self-interest will profoundly influence the future of conservation in America.

*
**

Small books can have big aims. Our goal is to provide a guide for how the conservation movement can effectively advance its agenda over the near and long-term future. We argue that understanding the clash of forces currently dividing the nation, responding to the emerging assault on the environment by the Trump administration, and framing conservation in new and transformative ways are together the best path for progress. We write this book for leaders of environmental organizations, professionals in conservation advocacy groups, federal, state, and local policy makers, resource managers, scientists, students, and interested citizen activists.

Our views regarding the future of conservation reflect our personal and professional paths. Raised in urban Seattle in the Pacific Northwest, Gary Machlis gravitated toward science, studied sociology and ecology, and has followed a career path of conservation research in U.S. national parks and forests, China (on the giant panda), the Galápagos Islands (on the impact of tourism), and in Africa, Europe, and the Caribbean. Jon Jarvis, who grew up in rural Virginia, studied biology and had a career in the National Park Service as park ranger, resource management specialist, park biologist, chief of natural and cultural resources, superintendent at several parks (from Craters of the Moon National Monument in Idaho to Wrangell-St. Elias National Park and

Preserve in Alaska), and regional director. Our two paths crossed frequently, as Machlis conducted research in parks managed by Jarvis. In 2009, Jarvis was nominated by President Obama to be the eighteenth director of the National Park Service and confirmed by the U.S. Senate. Standing near Jarvis at his formal swearing in, Machlis became his science advisor, the first science advisor to the director appointed in the National Park Service. Thus began our work together in Washington, DC, which spanned both terms of the Obama administration and continues with collaboration on this book.

It should not be a surprise that many (but not all) of the events, actions, conflicts, and challenges described in this book reflect the National Park Service; we ascribe to the writer's dictum: *write what you know*. But there are other reasons to use the National Park Service as an illustrative example. The park ranger is an iconic role in American life, respected in public service, and widely trusted. The National Park Service deals with issues from climate change to civil rights; protects both natural and cultural resources; and partners with other countries, states, large cities, and rural towns. It has responsibilities that cross seventeen time zones and includes programs from bear management in Alaskan wilderness to historic-preservation tax credits in Detroit. In many ways, the service confronts the full breadth of challenges facing conservation in America. And the national parks have played a pivotal and enduring role in the history of American conservation, as we describe in the next chapter. We should be clear: the National Park Service is a useful example but not the exemplar, and our focus is on the future of conservation in America *writ large*.

Throughout this book, we emphasize a core concept: *strategic intention*. Conservation actions (protecting a watershed, reforming a management policy, creating an educational program, organizing an event, and so forth) are most effective when they build the foundation and momentum for future actions. That is, strategy should select tactics that, when successful, lead the way for yet additional progress. An example is the National Park Service centennial, which was intentionally designed and conducted to create a new generation of park visitors, supporters, and advocates (more on that later). This theme of "intentionality" runs throughout our narrative.

We believe there are effective and tested strategies that can help chart the future of a broadened American conservation movement. We offer them as essential and practical tools and advocate for their use at local, state, and national levels. Also essential for progress is a new and unified vision of conservation. We call for the various branches of the conservation movement—some as traditional allies and others as new partners—to share common cause and work more closely together than ever before. In what follows, we describe this unified vision of conservation and how it can succeed.

This is an intensely personal book, borne of practical experience. Together we have accumulated over eighty years of service to conservation and are grateful for the opportunities, hard lessons, adventures (and misadventures), and even the crises we have experienced. There are innumerable and extraordinary individuals that have inspired us—from presidents to park visitors and from government employees to Nobel Laureate scientists and young students. The varied American landscape and the nation's sprawling and diverse history have demonstrated to us their restorative powers. While there is "rough water" ahead, we remain optimistic about and confident in the resilient future of conservation in America.

CHAPTER TWO

An Enduring Responsibility

In early 1915, the industrialist and owner of the Thorkildsen-Mather Borax Company Stephen T. Mather used his brilliant marketing skills and enthusiasm for conservation to convince a group of influential men to join him for a camping trip in the Sierra Nevada mountains. The "Mather Mountain Party" included writers for the *Saturday Evening Post*, the vice president of the Southern Pacific Railroad, the ranking Republican congressman on the House Appropriations Committee, the president of the New York Zoological Society, and the publisher of the Visalia, California, daily newspaper. Joining them was a professional photographer, several prominent attorneys and businessmen, California's state engineer, and Gilbert Grosvenor, director of the National Geographic Society. The group included one park ranger who served as horse packer and two cooks, Ty Sing (considered the best camp chef in the West) and his assistant.

For two weeks, the Mather party camped in alpine meadows, plunged into cold streams, reveled under a starlit night sky, and silently absorbed the stunning vistas. Cunningly, Mather let the mountains "do their magic." After the mountain travel and evening campfires, all present committed their publishing and political power to the preservation of the national parks.[1] The following year, the National Park

Service was established, with Mather as its first director. The new agency was assigned an enduring responsibility: "to conserve the scenery and the natural and historic objects and the wild life therein and to provide for the enjoyment of the same in such manner and by such means as will leave them unimpaired for the enjoyment of future generations."[2]

Our contemporary proposals for meeting this enduring responsibility extend beyond the National Park Service and are based on several premises. First, for over 150 years, a passion for conservation in America has thrived in periods of growth and persisted in times of challenge. Second, the most successful individual actions (such as Mather's artful camping trip) have strategic intentions that lead toward larger conservation goals. Third, there have been several generational transformations—key opportunities when a new generation of Americans comes of age and empowerment. And fourth, conservation actions are most effective when tested, experienced, refined, and shared.

*
**

Coming out of World War II, the national parks had been intermittently closed and often neglected; the parks required a restoration of public commitment to their care. In 1953, popular writer Bernard DeVoto even penned an article for *Harper's Magazine* titled, "Let's Close the National Parks." He sarcastically called for the park closures because they were so underfunded and their condition so deplorable that they were an embarrassment to the nation.

His caustic critique as well as congressional demands to improve park conditions drew the attention of National

Park Service director Conrad Wirth; in 1956, Wirth launched Mission 66. This was a ten-year branding, marketing, and construction effort intended to connect a new generation of Americans, especially returning World War II veterans, to the national parks. With peacetime prosperity, an emerging interstate highway system, and clever automobile company advertising ("See the USA in Your Chevrolet"), the white middle class visited in droves with the baby boom children in the backseat of the family automobile. They saw and experienced their country through the grandeur of parks such as the Grand Canyon, Mesa Verde, Rocky Mountain, Yosemite, and Zion. With segregation starting to weaken as the civil rights movement converted protest to progress, and the nation's archaic Jim Crow laws slowly being eliminated, a small but emerging subset of black middle-class visitors also connected with the parks.

Mission 66 was an initiative intentionally aimed at the average American, inviting them for an affordable holiday, refreshing (or creating) their connections to nature and American history, and, through their visits, connecting them to the shared values of the nation. As public facilities modernized and the number of visitors swelled, new models of conservation such as land trusts and private conservation easements emerged. Conservation philanthropy rose not only as financial support for conservation organizations but also by donation of private lands to federal and state agencies for long-term stewardship.[3] Increased federal funding for land management agencies was bipartisan and robust, facilitating the expansion of protected areas, boundaries, campgrounds, visitor centers, and trails.

Mission 66 had another critically important effect. Many of the baby boom generation born between 1946 and 1961 experienced the parks firsthand as children, and some became environmental professionals, public advocates (creating the first Earth Day in 1970), environmental entrepreneurs (such as the founders of Patagonia or Royal Robbins), and political supporters of conservation. President Nixon, hardly an outdoorsman, accurately gauged this growing, broad public support for conservation and responded. In a remarkable special message to Congress, he described his conservation agenda: "This is the environmental awakening. It marks a sensitivity of the American spirit and a new maturity of American public life. It is a working revolution in values, as commitment to responsible partnership with nature replaces cavalier assumptions that we can play God with our surroundings and survive. It is leading to broad reforms in action, as individuals, corporations, government, and civic groups mobilize to conserve resources, to control pollution, to anticipate and prevent emerging environmental problems, to manage the land more wisely, and to preserve wildness."[4]

During this period of progress, Congress passed the bedrock of conservation legislation, including the Wilderness Act (1964), Land and Water Conservation Act (1964), National Historic Preservation Act (1966), Endangered Species Preservation Act (1966 and amended as the Endangered Species Act in 1969), Wild and Scenic Rivers Act (1968), National Environmental Policy Act (1970), and Clean Air Act (1970). These foundational laws undergird contemporary conservation policy and practice by federal and state resource agencies, provide the bases for decisions in the courts, and empower the national commitment to meet the recurrent responsibility of conservation.

*
**

The groundbreaking legislation, emerging new environmental organizations, heightened awareness of the earth's finite resources, and expanding advocacy efforts were not focused solely on the National Park Service. While the National Park System preserves spectacular scenery and recreational opportunities, harbors great biological diversity, and preserves much of our cultural history, it does not, of course, sustain conservation on its own. At the federal level, the U.S. Fish and Wildlife Service Refuge System fills an essential role in preserving specific habitats for many species and providing nature experiences to the public. The U.S. Forest Service's National Wilderness System offers wild places for solitude and outdoor recreation. The National Landscape Conservation System of the Bureau of Land Management protects extraordinary diversity found within Western deserts and rangelands. Many state, regional, and urban park systems offer refuge for migratory birds, protection of valued cultural resources, and places for the public to connect with nature and history near home. Nonprofit organizations such as The Nature Conservancy purchase, hold, and transfer lands to support the land's long-term preservation; local advocacy groups are watchdogs for environmental protection. Together, these conservation institutions contribute to resource stewardship across America.

This enduring responsibility for resource stewardship has evolved over time, based on public engagement, growing visitor use, new scientific knowledge, political trends, professional expertise, fluctuating funding, and more. The core criteria of conservation—to provide for the present and preserve for the future—has expanded to include elements of historic preservation, sustainability, renewable energy, climate change adaptation, transformative visitor experiences, wildlife management, environmental justice, gateway community economics, landscape-level connectivity, and ecosystem science. In many ways, the conservation movement is poised to flourish and expand. But there are barriers.

Current and future resource stewardship requires that past arbitrary boundaries between natural and cultural resource management be abandoned; nature and culture are inextricably linked. For example, American bison (*Bison bison*) are a natural resource, and as "buffalo" they carry a heavy mantle of cultural meaning to Native Americans and serve as a symbol of the American West (and, since 2016, our officially designated National Mammal). This broad and inclusive resource stewardship is of paramount importance and necessarily includes both an ethic of preservation *and* the technical skill and expertise to make and implement wise decisions.

Because conservation decisions have important consequences for resources, people, and communities, they are often contentious and sometimes highly controversial. We believe conservation decisions will be the most defensible, sustainable, strategic, and enduring when they are based on three criteria: the best available sound science, accurate fidelity to the law, and careful consideration of long-term public interest over multiple generations. It is admittedly a lofty standard, especially when merged with strategic intention.

*
**

Disturbing indicators of a weakening commitment to the enduring responsibility began to emerge in the 2000s: visitation to the national parks had been flat for a decade and traditional conservation organizations were struggling to attract new members who represented the changing demographics of the nation. The baby boom generation was aging and edging toward retirement. The younger, millennial generation was less connected to the outdoors. At the political extreme, there were legislative proposals for divesture of the public lands, the de-designation of national parks and the rollback of environmental laws such as the Endangered Species Act. These worrisome trends represented major challenges and, if left unchecked, could harm the future of conservation in America.[5]

The election of Barack Obama in 2008 brought to Washington, DC, the first African American president, who entered the White House with a message of hope and an audacious ambition to accomplish transformational change. The opportunity provided by the new Obama administration was well-timed to strategically refresh the commitment to conservation in America. The National Park Service centennial, if executed with strategic intent, could serve as a public platform and alarm bell to engage new audiences, frame climate change adaptation as an urgent challenge, and build the scholarly and scientific evidence to address a less predictable future.

The term "park" was interpreted broadly to encompass national, state, and local parks, other public lands, historic sites, and green spaces. Centennial components were designed so that a wide range of entities—such as land management agencies, nonprofit organizations, education institutions, scientific societies, and youth groups—could implement and embrace them.[6] The centennial served to foster the next generation of conservation advocates and, in doing so, test a set of conservation tactics and actions.

*
**

We knew that the investment in conservation that had the greatest potential for long-term return would involve focusing on youth and connecting with them through their families, schools, and social media. We also knew there were significant obstacles to overcome. The rapid rise in smart phones, tablets, and other technological conveniences require wireless connectivity not readily available in remote natural areas that are characteristic of many public lands. Authors such as Richard Louv, in his book *Last Child in the Woods*, have lamented that the digital experience is replacing the actual experience.[7] Because of increased standardized testing in education, teachers spend an inordinate amount of time "teaching to the test," and their own experiences with the natural world and historic sites are often limited. Extracurricular activities like field trips have been curtailed due to severely reduced budgets. Many existing youth-serving organizations with emphases on outdoor experience and conservation projects for high school– and college-age students were struggling to attract and retain minority youth. Driven by the Great Recession that began in 2007, the number of children living in poverty had risen to 14.5 million—19.7 percent of America's children. Extreme poverty confronted one in eleven children.[8]

Approaches to connecting with youth had to be layered, comprehensive, and designed to meet their real needs. Taking inspiration from President Franklin Roosevelt's Civilian Conservation Corps, Secretary of the Interior Ken Salazar launched the 21st Century Conservation Service Corps, offering young persons from all backgrounds the opportunity to work in the outdoors and contribute to their country. By 2014, over twenty thousand youth from all fifty states had participated.[9] Time will tell if their experiences translate into lifelong interest in conservation.

Millennials became a primary focus of the centennial celebration. The millennial generation is the largest and most diverse demographic in our history. They will inherit the nation's economy, institutions, landscape, laws, values, and priorities, deciding which to sustain and which to discard. They are the future advocates for conservation in America. To reach them, a public awareness campaign was launched, designed to penetrate the cacophony of social media entertainment and digital communication that fill their busy lives. A major advertising campaign began in 2015 with the introduction of Find Your Park. Funded entirely by record levels of philanthropic contributions to the National Park Foundation and supported by the marketing expertise of corporate partners, Find Your Park appeared across the nation, from a Times Square "takeover" to trending on Twitter. Media coverage and survey polling suggested the campaign captured the attention and interest of the millennials and others.[10]

*
**

We argue that it is critical to the future of conservation to make the movement relevant to all Americans, especially the more diverse millennial generation. "Relevancy" is not a politically correct slogan or a glib statement for media consumption. It is the foundation for building enduring support for conservation. We define relevancy as when an individual, family, or group makes a personal connection with a physical place that evokes a potent emotional response. At least two kinds of relevancy are essential to conservation—the connection to nature and the connection to history.

Relevancy can be created through a sense of awe when viewing a vast and unique landscape or the sense of peace that comes from a walk in a quiet, forested urban park. The science is incomplete on the relationship between nature and human psychology, but there is increasing interest in the social and health benefits of contact with natural settings. We do know that the effects of contact with nature can include enhanced curiosity about the natural world. For some, early contact with nature leads to a passion for its protection, the very roots of conservation. Henry David Thoreau (one of the first to call for a system of national parks), John Muir, Theodore Roosevelt, George Washington Carver, Rachel Carson, Bob Marshall, E. O. Wilson, and many others started their lives of conservation advocacy by exploring nature near home.[11]

The second and equally important foundation of relevancy involves making deep connection to sites of American historical significance. Public schools and institutions, museums, and local and national historical societies are all tasked with telling parts of the American story. Much is well told, and places such as Independence Hall or the Statue of Liberty engage millions of Americans with the nation's past and present. Yet the contributions of women and people of color have repeatedly been left out of the American narrative; textbooks and exhibits too often emphasize history as written by the victors or dominant culture.

The National Park Service initiated (with strategic intention) scholarly studies to document that the American historical narrative has significant gaps. Historians, writers, activists, and communities described the experience of and contributions by Latinos, women, Asian American and Pacific Islanders, and the lesbian, gay, bisexual, and transgender communities, seeking out the places and stories that fill those gaps. With strong and often bipartisan community support, President Obama established new national monuments that represented these untold stories: César E. Chávez, Harriet Tubman, Pullman, Belmont-Paul Women's Equality, Fort Monroe, First State, Honouliuli, Charles Young Buffalo Soldiers, Stonewall, Reconstruction Era, Freedom Riders, and Birmingham Civil Rights. These sites contribute to a more complete history of America and, by placing them alongside the Lincoln Memorial, Gettysburg National Military Park, and the Grand Canyon, help engage citizens who may feel distanced from the traditional and narrower American narrative. Such relevancy builds support not only for these historical places and stories but also for conservation in general.

*
**

Strategic intention was also aimed at the most dire existential threat of our age. The scientific community overwhelmingly shares consensus that (1) the planet is warming, (2) this warming is accelerated due to human activity, (3) warming is already having significant impacts, and (4) in the absence of immediate and sustained action, global warming will have extraordinary and harsh consequences for humankind. Climate change is the greatest threat to the integrity of all public lands, from wildlife refuges to national forests to urban park systems and marine protected areas. The future of conservation will unfold during an era of climate change, and the millennial generation will be grappling with the cascading consequences of climate change for their entire lives.

For the last half of the twentieth century and the beginning of the twenty-first, the management paradigm for the conservation of large parks and wilderness areas was to "let nature rule," return valued species that may have been lost due to human activity (e.g., wolves), restore natural processes such as fire, control the arrival of nonnative species, and prevent pollution and boundary encroachment. If all done well, this could meet the standard of "unimpaired." The 1963 report *Wildlife Management in the National Parks*, prepared by A. Starker Leopold (son of ecologist and conservationist Aldo Leopold) provided a guiding philosophy. Yet much has changed since the 1960s. We now know that the paradigm of protection and restoration that has guided management of parks and public for the last fifty years is no longer fully viable in an era of climate change.

A science committee of the National Park Service Advisory Board was tasked with reconsidering the older paradigm. This group of extraordinary scientists, including Nobel Prize winners, members of the National Academies of Science, and others, worked to carefully consider and make recommendations. Their resulting report, *Revisiting Leopold: Resource Stewardship in the National Parks*, ambitiously stated: "The overarching goal of NPS resource management should be to steward NPS resources for continuous change that is not yet fully understood, in order to preserve ecological integrity and cultural and historical authenticity, provide visitors with transformative experiences, and form the core of a national conservation land- and seascape."[12]

With *Revisiting Leopold* as source and inspiration, the National Park Service began the long and arduous process of transforming an advisory group's recommendations into formal policy. Technical groups were formed, meetings convened, consultations held, and drafts prepared. National Park Service employees were encouraged to comment and suggest improvements. Public review (some opponents predictably argued that too short a review period was provided) and final review by National Park Service leadership led to the completion and signing of the new policy as Director's Order #100 in December 2016, the last month of the centennial. (The Trump administration has rescinded this policy.) Over time, this new paradigm may well transform how resource stewardship is practiced in the national parks and other conservation areas.

*
**

Strategies such as celebrating the centennial, reaching out to millennials, expanding relevancy, and responding to climate change are vital to conservation progress, but they do not represent the full range of possibilities for strategic intention. Inevitably, unexpected events intervene. On April 20, 2010, the *Deepwater Horizon* drilling platform exploded and later collapsed into the sea, killing eleven men and spilling over 4.9 million barrels of oil into the Gulf of Mexico. It soon became one of the worst man-made environmental disasters in U.S. history. It was a reminder of the fragility of the planet, our national thirst for fossil fuels regardless of risk, the limits to technology and corporate self-oversight, and the need for better science to address the cascading effects of such disasters.

The spill and its response diverted significant attention, resources, time, and staff to the Gulf. It also provided the opportunity to create new ways to apply science during a crisis. Traditionally, federal Incident Command Systems for oil spills primarily employed *tactical* science—assessing weather, ocean currents, oil chemistry and dispersal, techniques for stopping the spill, and so forth. *Strategic* science that focused on immediate, midterm, and long-term effects of the crisis was not a core part of response, nor did decision makers have access to such information quickly or directly. With colleagues in the U.S. Geological Survey, we devel-

oped a scientific working group and organizational struc-
ture, modeled after the Office of Strategic Services during
World War II, and titled the new organization the Strategic Sciences Working Group. The working group delivered
important results to the Incident Command and resource
agencies and ultimately published its scenario work in
scientific journals. The group's findings also helped guide
priorities for the initial and subsequent British Petroleum–
funded multibillion-dollar restoration effort.[13]

Then, in October 2012, Hurricane Sandy advanced
along the Eastern Seaboard of the United States, making
landfall near Atlantic City, New Jersey. Combining with a
nor'easter, the storm affected seventeen states and caused
massive damage in both New York City and much of the
nearby coast. Hurricane Sandy presented an opportunity
for the now formal Strategic Sciences Group to assess the
effects of future climate-driven sea level rise, storm surge,
and impacts on coastal communities over twenty-five- and
fifty-year time horizons—and again help guide federal restoration and response. The establishment of the Strategic
Sciences Group demonstrates the potential to apply strategic intent even when there is a crisis.

Less unexpected but still a detour from the path of intentional progress, the sixteen-day federal government shutdown in 2013 became an exercise in political theater, diver-
sion of administrative effort, and economic disruption. The
public outcry over the closure of parks ultimately forced the
hand of Congress but reaffirmed the strong position that
public lands have in American society. The National Park
Service director (Jarvis) was grilled before a joint hearing

of the House of Representatives, defending the requirement that parks need federal funding in order to provide for their protection and for visitor enjoyment. The federal government eventually reopened, as did national parks and wildlife refuges. The shutdown and startup of the National Park System is neither simple nor elegant; visitors are forced to abandon long-awaited experiences in parks and even in-park weddings are cancelled. The return to normal operations absorbed much of the following month.

A serious and substantial problem emerged during this period, as a high-profile case of sexual harassment by rangers at Grand Canyon National Park led to broader revelations of harassment at other national parks and in other federal resource agencies. The results were acrid congressional hearings, widespread media attention, and a growing realization (some would say too slow to emerge) among agency leadership that the corrosive impacts of sexual as well as other forms of harassment could not be ignored. In the final months of the centennial, the National Park Service and other agencies focused inward and on responding to the reality of workplace harassment. A deep and critical assessment has begun, as well as initial steps at corrective action to create a more respectful agency culture and remove the harassers from federal employment.

Many of the opportunities and crises just described had an impact on other conservation organizations. The U.S. Fish and Wildlife Service and the National Oceanic and Atmospheric Administration documented the harm to birds and other wildlife from the response to the *Deepwater Horizon* oil spill. Departments of natural resources from the Gulf States, nonprofit organizations, regional universities, research institutions, and city governments all engaged actively in response and recovery efforts. Many states and private-sector companies joined in celebrating the National Park Service centennial. For example, the Massachusetts Department of Conservation and Recreation launched a series of "centennial" hikes in 2016 to encourage public use of their state parks. Urban initiatives paired federal agencies and cities such as Detroit, Miami, and Denver to create walking paths, urban wildlife refuges, and reconnections to improved waters of working rivers.

The actions of strategic intent described above were gaining momentum and making progress when, toward the end of the centennial year, the 2016 presidential election was held. Conservation was dealt a stunning blow.

CHAPTER THREE

A Chart for Rough Water

In 1940, writer and activist Waldo Frank penned a prescient book on the future of Western democracies, titled *Chart for Rough Water*. Frank argued that, after the democracies defeated Nazi Germany and Imperial Japan, they would have to turn to the long war against communist totalitarianism, and when it was defeated (or at least contained), turn finally to their internal divisions of racism, inequality, and misuse of resources. The book was reprinted in 1947 along with his earlier call for a renaissance of national spirit in *The Re-Discovery of America*.[1]

Frank (who served as an envoy for President Franklin Roosevelt) warned that Americans must confront immediate challenges that "clamor frantically to be solved" yet, at the same time, address underlying causes that will require progress that "fills the lives of even our youngest children." President Roosevelt understood this: his New Deal programs such as the Civilian Conservation Corps, which introduced millions of young men to the outdoors, incorporated immediate jobs for the unemployed, social security for all citizens, and long-term conservation of natural resources. This expanded and intergenerational view remains relevant in our era of rapid climate change, globalization, extreme inequality, and national civil discord. The long view is vital to the future of conservation in America.

With the onset of the Trump administration, the nation and the conservation movement must now navigate some very "rough water." As the outlines of the assault on conservation emerge, there is a rising sense of dread and concern, as well as a search for guidance. We necessarily pivot from the National Park Service as an illustrative example, to a more general question: What is ahead for conservation, environmental science, and protection of America's resources?

*
**

Regardless of political party, narrow self-interests, or even well-intentioned actions, there are major environmental and social threats that confront America. They form the underlying causes that frame our immediate conservation problems and will challenge this and future generations. Three of these threats can serve as interdependent examples: climate change, species extinction, and economic inequality.

Climate change is likely to accelerate given the inadequate response to date, and the first great wave of severe impacts such as environmental refugee migrations will soon be undeniable. The Intergovernmental Panel on Climate Change projects global surface temperatures to exceed 2.7 degrees Fahrenheit above preindustrial levels by the end of the century. The alarming cascade of consequences includes accelerated melting of the Greenland ice sheet leading to rapid sea level rise, along with increased release of carbon from the melting permafrost, which in turn amplifies the greenhouse effect and further increases climate change.[2]

Climate refugees already exist in the United States: Isle

de Jean Charles in Louisiana is the first in the United States to receive federal funding to resettle an entire community. It will not be the last. The Union of Concerned Scientists warns: "Americans in some communities already know what it feels like for the slow creep of sea level rise to intrude in some way on their daily lives, flooding their neighborhood or place of work, rerouting their commute, driving down the value of their home. In the decades ahead, though, many more of us will experience these changes." In the absence of preventive actions, by 2035 nearly 170 coastal U.S. communities will reach "chronic inundation"— flooding, on average, every other week. These communities will face profound decisions and harsh alternatives of "defending against the sea, accommodating rising water [or] retreating from flood-prone areas." Rising sea levels may permanently flood not just single communities but hundreds of U.S. counties; one of the most threatened is Miami-Dade County in Florida, putting almost two million persons at risk. In his book *The Southern Diaspora*, J. N. Gregory reviews the scientific assessments and warns: "These results suggest that the absence of protective measures could led to US population movements of a magnitude similar to the twentieth century Great Migration of southern African-Americans."[3]

The volatility of floods, fires, droughts, and storms will strain the nation's emergency response and restoration capacity and make "rough water" more than a mere metaphor. Global climate change has direct effects on regional climate patterns and heightens the extremes of weather, such as seasonal periods of extreme heat days, or "heat waves." The U.S. Global Change Research Program's third national assessment makes this ominous comparison: "Climate models project that the same summertime temperatures that ranked among the hottest 5% in 1950–1979 will occur at least 70% of the time by 2035–2064."[4] In human terms, what was a rare extremely hot day for baby boomers as teenagers will be commonplace for the millennials in their middle age. Similarly, the risk of catastrophic hurricanes is projected to almost double, and sea level rise combined with intensified storm severity will create even higher flood risk for coastal areas. These events will stress America's already vulnerable infrastructure, "including power, water, transportation, and communication systems that are essential to maintaining access to health care and emergency response services and safeguarding human health."[5]

Similar to the climate change threat (and linked to it by numerous pathways), species decline and extinction rates will likely accelerate. The long view suggests that the earth is in the midst of a modern, human-caused "sixth extinction" documented in Elizabeth Kolbert's Pulitzer Prize–winning book *The Sixth Extinction: An Unnatural History.*[6] A 2017 study published in the *Proceedings of the National Academy of Science* found that species extinction is accompanied by devastating declines in vertebrate species populations and habitable ranges. The study's authors estimate that 22 percent of the species in North America have experienced range contractions of at least 80 percent. They warn: "In combination, these assaults are causing a vast reduction of the fauna and flora of our planet. The resulting biological annihilation obviously will also have serious ecological, economic, and social consequences."[7]

As species such as honeybees or even entire fisheries disappear, some ecosystems will unravel and destabilize, and economic disruption (particularly in agricultural production) will increase. Pollination, crop yield, and livestock production will be stressed and likely reduced, leading to less yield and higher prices. Already, the cost of food and lack of access to healthy food now affects forty-eight million Americans.[8] Without significant improvements in food security and availability, more Americans will go hungry in the future. Clearly, climate change and species loss have social and economic consequences.

Economic inequality will likely increase, driven by the concentration of wealth in the hands of those that set economic policy and derive profits from these policies and exacerbated by the disruptions caused by environmental stress on the national economy and local communities. Evidence and cautious projections suggest that climate change under even conservative models of warming will increase preexisting inequality among regions of the United States, with southern and midwestern populations suffering the largest losses and northern and western populations facing smaller losses or even net economic gains.[9]

At the national level, inequality has become grave, chronic, and deeply divisive. According to Inequality.org, a project for the Institute of Policy Studies, "the gap between the rich and everyone else has been growing markedly, by every major statistical measure, for some 30 years." America's top 10 percent average an income nearly nine times as much as the income of the bottom 90 percent, while the top 1 percent averages an income over thirty-eight times more than the bottom 90 percent. Staggeringly, the nation's top 0.1 percent make an income on average of over 184 times the bottom 90 percent.[10]

This widening wealth gap threatens mobility, education, food security, and access to health care. It also damages our collective civic life. Sociologist Thomas Shapiro argues that wealth disparities are closely tied to racial inequities, in what he describes as a dangerously combined "toxic inequality."[11] The challenge that economic inequality represents in the long view is deeply structural, eroding core values of national solidarity and threatening "functional citizenship" for large numbers of Americans. Political philosopher and life-long activist Noam Chomsky, writing about the concentration of wealth and power, is bitterly direct: he titled his 2017 book and film *Requiem for the American Dream*.[12]

*
**

While the underlying causes (those described and many others) will challenge conservation now and in the long term, the impact of the 2016 election will have immediate and damaging consequences.

First, there has been and will continue to be a clawback of constructive response to climate change. As Amanda Erickson wrote in the *Washington Post*, "climate change denial is not incidental to a nationalist, populist agenda, it's central to it." Because climate change is a global problem and international cooperation is a prerequisite for action, climate change is an inconvenient truth that conflicts with the "America First" worldview.[13] Climate change denial also feeds the populist anger at elites and experts. Tom Nichols, professor of national security affairs at the U.S. Naval War College has noted: "To reject the advice of experts is to assert autonomy, a way for Americans to demonstrate their independence from nefarious elites—and insulate their increasingly fragile egos from ever being told they're wrong."[14]

The climate change reversals begin with the legitimizing of climate-change deniers—and appointing them to positions of authority, such as Scott Pruitt as administrator of the Environmental Protection Agency. Over time, positions as deputy directors, advisors, counselors, program leaders, and others will be filled with climate change skeptics. They are likely to lead the steady retreat of federal activity from proactive response to climate change impacts—even as these impacts become more pronounced. The official strategy will be to deny and denounce future projections.

From these positions of federal authority, climate science (and the critical data it generates), mitigation programs, and international commitments will be scrutinized, and the Trump administration will seek to abandon them, in some cases succeeding (such as the Paris Accord). Suppression of climate science has been and will continue to be quick and severe—selected websites (such as those on Arctic climate change) were deleted within days of Trump's inauguration, and more continue to go dark. Climate change research will be targeted for reduction or elimination—an effort to stop the flow of facts at the source. Essential programs such as NASA earth sciences research, beaming critical data from its satellites back to Earth, will be threatened. Attempts will be made to purge long-term data sets, eliminate field stations, and shutter climate change research centers. The intentional interruption of data flow has cascading and deleterious effects on new scientific understanding, evaluating management actions for effectiveness, developing sound policy alternatives, and smart decision making. Even modest gaps in time-series data sets (such as the timing of migratory movements of birds or snowpack measurements) can substantially reduce the value and robustness of the data; that is the intent, of course.

The abandonment of national climate change actions will necessarily be incremental and haphazard, as they have been established across a variety of federal agencies and their cancellation or reduction may require congressional approval. Just in the spring of 2017 alone, the administration's Department of Transportation began rolling back new fuel-economy standards for cars, the Environmental Protection Agency rejected President Obama's Clean Power Plan, and President Trump theatrically announced withdrawal from the Paris Accord, the major international agreement on greenhouse gas emissions and climate change mitigation. Not all of the attempted climate change reversals will succeed—there are substantial economic, social, and even national security interests that will push back, along with many states. But even though climate change is an underlying cause creating fundamental transitions in the national economy, environment, and society, its denial will remain a core strategy of the populist movement.

Second, there will likely be a retreat from large landscape conservation—that is, the need to assemble and co-manage lands and waters to reflect the landscape-level requirements of wildlife, forests, and other ecosystems. Carefully built partnerships between federal agencies, state governments, and local resource managers, such as the successful Great Lakes Restoration Initiative and the emerging Sage Grouse Conservation Plan, will be threatened with defunding or termination. Conservation actions undertaken by the Obama administration, such as its active use of the Antiquities Act, will be targeted for reversal or reduction. A test case is likely to be Bear's Ears National Monument, given its size (1,351,849 acres), location (the conservative state of Utah), and the innovative engagement of the Hopi Tribe, Navajo Nation, Ute Mountain Ute Tribe, Ute Indian Tribe, and Pueblo of Zuni. Policies carefully crafted to meet the challenges of the twenty-first century, such as those emanating from the report *Revisiting Leopold* described in chapter 2, will be challenged, with the intent to rescind them in the vain hope that out-of-date twentieth-century policies will protect special economic interests from the need to adapt and change.

The attacks are already being characterized as a defense of "states' rights," "private property rights," and opposition to "federal government overreach." Agencies will be silenced to prevent "mission creep" (in reality, collaboration), and interagency task forces, working groups, advisory boards, and consultations will be suspended, weakened by neglect or direct elimination, or ignored. Some of the more extreme proposals, such as moving the Department of the Interior headquarters from the nation's capital to somewhere in the western United States or transferring significant amounts of federal land to the states, will likely be rejected but will nevertheless require opposition groups to expend precious time and resources countering even the most obviously injurious of proposals.

Third, there will be the erosion of science (not just climate science) from its proper role of informing policy to the marginalized position of "organizational pest." Coming is a kind of American Harperism—Canada's experience with the silencing of environmental science by pinpoint budget cuts and removal of scientific advice from decision making.[15] The erosion begins at the White House, which (as of mid-2017) has neither nominated nor appointed an individual to the position of science advisor to the president, a position first established by President Franklin Roosevelt in 1941 and filled by every president since. The need for the Office of Science and Technology Policy and the science-dependent Council on Environmental Quality has been questioned if not outright mocked by Trump advisors, and will likely be reduced in scale, scope, and influence.

This attack on science is not unprecedented; a metaphor of violence is expressed in books such as *Science under Siege* (1998), *The Republican War on Science* (2005), and *The War on Science* (2016).[16] The midterm outlook is grim. Ongoing science will be strategically assaulted by discontinuing inventory and monitoring programs that collect and manage crucial long-term data sets, eliminating scientists' positions on the grounds of budget constraints, and muzzling outspoken science professionals by rule, travel restriction, transfer to nonscience positions, and outright censorship. Talented scientists will depart in frustration, and then, when the programs are weakened, those remaining will face threats of elimination based on "poor performance," "accountability," or "cost cutting." Science integrity offices established during the last administration will be stripped of responsibility and authority, freeing policy makers from restrictions enacted to combat misconduct and misuse of science. Science advisor positions will be either left empty or staffed by politicized professionals, and scientific advisory groups will be eliminated or filled with pseudoscientists and special interest representatives adhering to the administration's strategy of weakening science.

The irresponsibility in denying climate change and the dereliction of duty to respond to its consequences, the dim understanding of threats caused by species decline and extinction, a toxic disregard for rising inequality, the retreat from large-landscape conservation, and the sustained attack on science now underway will continue throughout the Trump administration and perhaps beyond. All of these actions and inactions can deeply hinder and harm conservation. Truly, there is rough water ahead.

CHAPTER FOUR

Strategies for the Future of Conservation

As the wave of populist anger and resentment bears down on the national laws and policies that protect the environment, there is an inclination in the conservation community to veer haphazardly from cautious hope to indecisive despair. Cheery optimism that the rough water ahead is the briefest of storms or that it can be willfully ignored is ill-advised. So is succumbing to a debilitating anxiety about the future. The former prime minister of the United Kingdom, Tony Blair, reminds us: "Outrage is easy: strategy is hard. Outrage provides the motivation. But only strategy can deliver victory."[1] We would add that strategy must be combined with a suite of specific skills and attributes: adherence to principle, strategic intention, experience, willingness to learn, fortitude, collaboration, and perseverance.

The founding principle for modern American conservation comes from the Founding Fathers and their declaration that the core birthright of Americans is "life, liberty, and the pursuit of happiness." To be clear: a clean, healthy, and sustainable environment, anchored by parcels of lands and waters protected from despoliation and able to be enjoyed by all citizens unimpaired for future generations is an essential part of the American birthright. To the extent that conservation adheres to this principle, conservation is at core patriotic, bipartisan, relevant, and sustainable.

In this chapter, we suggest a series of important conservation strategies that national, state, and local leaders, government employees, corporate executives, advocacy group members, scientists, students, and committed citizens might consider and pursue. Some require immediate action. Others may require continuous effort over the course of several years or even decades—the long view of conservation. Some of these strategies have been familiar for years but sporadically practiced. Many are already emerging as actions by current conservationists. Several strategies we recommend have yet to be embraced. We believe all are important.

*
**

Monitor, record, and expose the retreat and retrenchment of environmental protection and conservation

Conservationists need to document and share widely every act of suppression, elimination, and revision of regulations and policy by the Trump administration. Advocacy groups should establish fast-responding websites, social media strategies, and annual reporting of exactly where and how the administration and the federal agencies under its control abandon environmental protection in America. One model is Evidence for Democracy, the Canadian advocacy group that battled the Harper government and remains an influential watchdog organization. The accurate and fair documentation of any and all retrenchment is a critical protection of democratically achieved conservation gains and a necessary foundation for future action.

Organize and act to provide science in the public interest

As advocacy groups ramp up their countermeasures, including public education, political advocacy, litigation, and more, the need for scientific information will exponentially increase. Data havens (each a refuge to protect threatened data) have begun to appear, often managed by scientific organizations, universities, or advocacy groups. These and other means to protect information should be encouraged, provided with necessary resources, and stocked with legitimate and critical data. Most vulnerable are long-term monitoring data collected by the federal government but still currently accessible. Many of the fiercest struggles will come in the courts and Congress, where some evidence-based decision making will still take place. The protected data will be invaluable in supporting those decisions.

Not all advocacy groups can afford scientific staff, and hence, science for the public interest must be constructed by the broader scientific community. This includes citizen science—which, if conducted carefully, can extend the reach of data collection and further public engagement. There are examples: ranchers are monitoring air quality around fracking wells, the citizens of Flint, Michigan, are testing for water safety, and Trout Unlimited is monitoring stream health, all under the guidance of professional scientists based in universities. In addition, scientists need training in giving effective testimony, and "science for litigation" should become an important mission for many scientific organizations, as well as a tactical response by advocacy groups.

Base the communication of conservation on deeper understandings and broader coalitions of interest

A broadened communication strategy for conservation is essential. It is not the simple creation of hashtag monikers, celebrity-infused public service announcements, or booking more conservationists on talk shows friendly or hostile to the conservation agenda. A more robust communication strategy requires the difficult examination of underlying causes. In 2015, the Pew Foundation completed a major study of public and scientific community attitudes toward science. The results were revealing: the general public does not share the scientific consensus about evolution. While 87 percent of scientists believe life has evolved over time owing to natural processes, only 32 percent of the general public shares this belief.[2] This should concern every conservation activist and scientist.

Whether the subject is evolution, global biodiversity, regional threats caused by climate change, or the urgency of local environmental protection, simply communicating *more* without understanding the deeply held values of the American public is not likely to be successful. In her 2016 book *Strangers in Their Own Land*, emeritus professor of sociology Arlie Hochschild describes her struggle to understand "why some Americans who may be most in need of government protections seem most hostile to them."[3] It is a struggle the conservation community must join, as an immediate step toward long-term progress in communicating the importance of American conservation. Fieldwork, sustained listening, and unbiased public surveys (not the "surveys" distributed by advocacy groups to raise money or increase membership) should be used to inform serious reflection and redirected effort. A broader coalition of park rangers, classroom teachers, field naturalists, museum docents and volunteers, tour guides, youth group leaders, popular writers, filmmakers, enlightened corporate executives, and religious leaders can and should play a strong role in an ongoing dialogue with the public on conservation issues. An example is Pope Francis's 2015 *Encyclical on Climate Change and Inequality: On Care of Our Common Home* and its powerful message not only to Catholics but to all persons.[4]

Incorporate strategic intent and the long view into every action, conflict, or crisis

Every day there are events, media stories, court proceedings, environmental crises, political decisions, and individual and agency actions affecting the American landscape. While these individual occurrences may seem disconnected or episodic, each presents an opportunity to apply strategic intent to bolster the future of conservation. Too often, conservation advocacy organizations stumble from each defensive action to the next or tout singular achievements (however worthy), focusing on relatively short-term wins and losses. Incorporating strategic intent into conservation action is not simply the creation of a vision statement, annual plan, or other organizational device. These are certainly necessary, but what is crucial is the commitment to identify how each individual action in the present can contribute to conservation in the long view. The future of conservation requires that leaders devote skill, time, energy, and funds to well-planned and intentional actions for the long view—actions that create critical momentum for future advances.

Every occurrence, however devastating, can present a juncture for action. Hurricanes can result in human and economic tragedy, yet their aftermath creates an opening for public conversations about using conservation practices (such as "green infrastructure") to mitigate future storms, sea level rise, and climate change. Popular writing on the scientific contributions of women and minorities (as recently chronicled in bestselling books and films such as *The Immortal Life of Henrietta Lacks* and *Hidden Figures*) reveal stories of discrimination that present fresh opportunities to build relevancy to new communities.[5] Innovative experiments in public education provide openings to integrate field experiences and conservation into childhoods and to engage parents as well. Each conservation struggle regardless of outcome carries a lesson that needs to be analyzed carefully, told accurately, and shared without malice, so as to inspire advocacy for the next conflict that will surely arise.

Connect all citizens to American nature, history, and culture through first-hand experience

Throughout this book, we have argued that first-hand experience with nature and history is a foundational element in developing conservation values and support. What was true for Franklin Roosevelt, Rachel Carson, the baby boomers who experienced Mission 66, and the millennial generation engaged with the National Park Service centennial is still true today: first-hand experience leads to awareness, appreciation, and advocacy.

An important strategy for enhancing first-hand experience is to better coordinate the activities of multiple institutions and to make each a portal of entry to the others. Our public education system, land management agencies, zoos, libraries, local parks, museums, and aquariums tend to operate in relative isolation, independently presenting their own stories and providing their own experiences, rather than collaborating in a larger learning narrative. We argue that this can and should change. Textbooks in subjects from history to biology can link their subjects to opportunities for live experiences—wherein reading about Revolutionary War history promotes a visit (real or virtual) to related historic sites. Zoos can link their exhibits to the public lands that preserve and offer access to the native habitats of zoo-displayed animals such as the American bison. Museums can encourage the visitor stirred by the newly established Smithsonian Museum of African American History in Washington, DC, to also experience and reflect on the newly established Fort Monroe National Monument in Norfolk, VA, where the first slaves were offloaded in America.

But encouraging first-hand experience is not enough; programs must be expanded that create real and affordable opportunities in order for those experiences to happen. Bold initiatives are called for. Priority should be given to youth programs, as an intentional strategy for building future support for conservation. These experiences can and should be transformative: they should educate *and* inspire. Many youth-serving organizations are very effective, including Nature Bridge, the Student Conservation Association, Outward Bound, and many state-based conservation corps, among others. These first-hand experiences can lead to a life connected to American nature, history, and culture, and hence, conservation.

Expand recognition and sharing of
the full American story

Every aspect of American history has more than one story, more than one perspective. The Americas were not "discovered" by Columbus, as there were millions of pre-1492 inhabitants with elaborate language, art, architecture, music, religion, and trade.[6] Westward expansion ultimately connected the West, midcontinent, and East Coast into a continental-scale country but decimated the existing nations of Native Americans. The Emancipation Proclamation granted freedom to four million African Americans, but they still had to fight for another hundred years to achieve legally protected civil rights. Hispanic heroes fought in the Civil War, suffragettes chained themselves to the White House fence, and Japanese Americans fought bravely in World War II even as their families were incarcerated in camps back home.

Expanding the recognition and awareness of the full story of America is essential for two reasons: vigilance and relevancy. Revisionist and selective history is the tool of despots and dictators, tailored to keep them in power and their followers in allegiance. The national narrative is always evolving, and its arc must bend toward a fuller truth. Sharing the entire and accurate story acts to protect and defend American ideals. Historians, scholars, writers, and students must seek out the multiple perspectives of our national history and ensure those perspectives are recognized and respected.

Engaging groups whose stories have not been told, told inaccurately, or told with misinformed condescension is a powerful strategy for building relevancy. The National Trust for Historic Preservation, State Historic Preservation Officers, and organizations like the Association for the Study of African American Life and History recognize that the current inventory of places of historical significance underrepresent the contributions of women and minorities. While much progress was made during the previous decade, this effort to enlarge the American narrative must be continued, well-funded, and converted into recognition and preservation of additional historic sites.

Extend the healthy parks, healthy people strategy to more Americans

There is growing scientific evidence that outdoor activity in nature is correlated with reductions in health risks including high blood pressure, cardiovascular disease, diabetes, and obesity. Activity in "green settings" (from small city parks to wilderness) has been shown to offer measurable mental health benefits and marked improvement in cases of depression, anxiety, psychoses, and post-traumatic stress disorder. As understanding of how our environment influences individual health grows, the evidence that supporting healthy parks (and by extension all public lands) leads to healthy people will accumulate.[7]

The next important step is to expand existing efforts to include more Americans. Outdoor recreation should be considered a healthy choice opportunity; hiking clubs and equipment companies can and should be part of American health care. Environmental education at parks should extend even further to include learning about local soils, plants, gardening, and nutrition. Where these programs are underway (and they are thriving in many parts of the country), they should be expanded. Additional efforts should target underrepresented communities, at-risk youth, single parent families, and both urban and rural vulnerable populations—all of whom suffer inequalities of health and unequitable access to health care. Leadership at neighborhood, city, and state levels will be essential and effective. Small city parks, trail systems, urban gardens, county and state parks, and other green environments need to be redefined and reconceived: park resources are *health* resources and can be managed to serve and contribute to American's better health and longevity.

Clinical training of general practice doctors, specialists, psychiatrists, clinical psychologists, and other health professions should include "environmental health care"— and perhaps college students in premed should, along with chemistry, be required to study ecology. As clinical evidence increasingly confirms that healthy parks mean healthy people, it should become standard practice for medical professionals to prescribe outdoor activities in nature to patients across America.

Protect, connect, and grow the network of protected areas across the American land- and seascape

The expansive network of protected areas in the United States includes over 25,800 local, county, state, private, and federal terrestrial parks protecting (at varying levels) 14 percent of the American land base. Data for marine protected areas varies widely by the methodology used; the advocacy group Our Ocean reports fully protected, no-take reserves account for 3 percent of U.S. waters.[8] While this represents a meaningful achievement, it is insufficient for several reasons.

Climate change and its ecological consequences require that existing protected areas be augmented, protecting key adjacent lands and waters to plan for range shifts and new seasonal migration patterns. In addition, there are essential lands and waters not yet protected, and certain ecosystem types (such as North American temperate grasslands or the floodplain forests of the Mississippi river deltas) are not fully represented in the current network. The accelerating loss of species and decline of populations (from large invertebrates to microorganisms) that are the foundation of life on our planet have devastating consequences for both nature and society. Visionary scientists such as E. O. Wilson and his Half Earth Project have articulated the threat and recommended crucial actions that can be taken to expand the protection of biodiversity.[9] These include significantly expanding the network of protected areas.

Creating physical connections among protected areas—"connectivity" in conservation science parlance—is an essential and strategic focus that conservation groups should prioritize. Corridors that follow river courses, abandoned railroad routes, and utility lines can link city parks, wetlands, open space, state parks, working forests, conservation easements, federal recreation areas, and wilderness. There are successful examples that can be used to support additional connectivity. The path of the pronghorn antelope (*Antilocapra americana*) is well known, and the pronghorn's seasonal route from Grand Teton National Park has been identified, protected, and enhanced. Even small sites can be beneficial. In Chicago, a local community restored a small area adjacent to an industrial railroad that had been filled with trash and abandoned vehicles, turning it into a park with native plants and a small wetland. Soon migrating Monarch butterflies began using it to feed and rest.[10]

Strategic expansion of protected areas should be a priority at every level, and advocates should build broad support beyond the traditional conservation community. Many of America's mayors are recognizing that parks, riverfronts and lakefronts, greenways, tree canopy, and other outdoor recreation sites attract new companies and are keys to a livable city. There is increasing evidence that green infrastructure contributes to crime reduction, public health, water quality, storm protection, reduced energy consumption and the social fabric of a community.[11] As a consequence, the public should demand greater support for and investment in local parks and green spaces. Additional discarded industrial sites adjacent to rivers and shorelines can be redeveloped and restored for recreational uses such as bicycling, community events, kayaking, and swimming. Long distance hike/bike routes along rail, pipeline, and energy corridors are being constructed throughout urban America, such as the ninety-four-mile rails-to-trails Silver Comet project in western Georgia and eastern Alabama; more should be encouraged. Linking these local investments across landscapes to larger parks, strategically bridging barriers both physical and economic, and securing the growing network of protected areas can and should be accelerated in the decades ahead.

Invest time, effort, and resources in
local engagement with respect for local values

All conservation is local. There is an untapped wellspring of conservation advocacy in places like the neighborhoods of Baltimore, Louisiana parishes, New England townships, the Canyon Country of Utah, New York City boroughs, the ethnic enclaves of Miami, and the rural counties of middle America. The durability of future conservation accomplishments will hinge on multigenerational local support. This requires renewed efforts at local engagement. National conservation groups periodically make two classic mistakes: underestimating the depth and value of local knowledge and/or assuming there is little local interest or support for conservation. National groups need to work collaboratively with local organizations of all types (not just conservation ones) to invest in responsive regional programs and knowledgeable field officers—hired locally whenever possible. Local values and indigenous voices must be incorporated into the formula for conservation; nontraditional conservationists should be cultivated and engaged; local leadership should be empowered.

It is also necessary to abandon the comfortable mythology that exists in many government agencies—assuming that open space and recreation services can wait for their clients to appear. These agencies must become proactive, dynamic, and creative organizations that reach out to serve previously underserved populations, expand services, and seek new constituents by establishing themselves as part of the local community's future.

A key element of local engagement is active and sustained listening. Conservation goals must include "green" jobs, recreational access, lifestyle choices, and traditional uses that are not only compatible with but also potentially enhance the conservation action. This requires time, effort, patience, and respect for local knowledge. It may not be easy or comfortable, but it *is* necessary. As Tom Snyder notes, "First, ideas about change must engage people of various backgrounds who do not agree about everything. Second, people must find themselves in places that are not their homes, and among groups that were not previously their friends."[12]

There are models. The nearly sustainable ecosystem of the Blackfoot Challenge collaboration in the northern Rocky Mountains would not have been possible without lengthy conversation and debate among fishers, hunters, foresters, ranchers, community members, and conservationists. President Obama's designation of Katahdin Woods and Waters National Monument would not have been achieved without the sustained listening and adjustments of strategy by Elliotsville Plantation president Lucas St. Clair, as he has said, "over a thousand cups of coffee." Perhaps sharing a thousand cups of coffee should become a key measure of effort by conservationists, and local engagement a reemphasized strategy for conservation action.

Increase investment in science and better integrate scientific information and insights into decision making

The most effective conservation decisions are those based on accurate understanding of key conditions. Science—including the biophysical and sociocultural sciences as well as interdisciplinary research—can provide the best available understanding as well as insight into the cascading effects of environmental actions. But to be effective, scientific research must be usable knowledge, "rigorous in method, mindful of limitations, peer reviewed, and delivered in ways that allow managers to apply its findings."[13] New disciplines such as paleobiology and cliodynamics (the quantitative study of long-term socioeconomic trends), along with new techniques using big data analytics and artificial intelligence hold much promise for conservation.

Such science must be unfettered by censorship and not starved for lack of funds. Investment in conservation science must be increased, and this investment cannot be restricted to the federal government. The funding stream for conservation science must be diversified. Large foundations like the Bill and Melinda Gates Foundation, Rockefeller Foundation, and Ford Foundation should expand their formidable portfolios to support increased research relevant to conservation. State agencies should prioritize within their already strained budgets the inventory and monitoring of environmental conditions most essential to decision making. Some corporations that extract profits for shareholders from outdoor recreation-related purchases currently devote a percentage of their resources to conservation science (such as the 1% for the Planet initiative), treating this cost as a research and development investment. More companies should follow their example.

New forms of science funding—ranging from small-scale crowdsourcing to large prize competitions—should also be supported. And because the federal government has special responsibilities and significant resources, the conservation community must continually press, persuade, cajole, and demand that federal agencies such as the Environmental Protection Agency, the National Park Service, and the U.S. Geological Survey (and their congressional appropriation committees) support active programs of relevant, applied science.

However, simply funding more science will not be effective if it is not paired with better integration of science into decision making. We do not suggest that all conservation decisions can or should be based on science alone; science-based decisions can often be unsustainable in the political minefields of conservation conflicts. Yet science-*informed* decisions are necessary and wise—whether it is winter-use limits in Yellowstone National Park or designing resilient infrastructure in New York City. Resource managers must increase their scientific literacy, and a foundational level of scientific literacy should become a prerequisite for those ambitious to become agency leaders. In addition, the scientific community and individual scientists have a responsibility to greatly improve their communication skills, as well as devise better ways to communicate uncertainty, limits to knowledge, and the potential economic and social consequences of environmental decisions.

Integrate climate change into all conservation decisions and actions

We have stressed in this book that climate change is the defining challenge of the present, as it will be for the foreseeable future. We do not adhere to the dark and dystopian scenarios of societal collapse. But we also do not underestimate the hardships, disruptions, and fundamental changes that anthropogenic climate change will bring to the planet, our nation, and our communities. Bob Dylan is right, as he is so often: "It's a hard rain's gonna fall."

Global greenhouse gas emissions must be reduced, and the conservation community should unequivocally support the Paris Accord provisions even as the Trump administration abrogates and abandons its responsibilities. States like California (with the sixth-largest economy in the world) should persevere in their efforts to mitigate and adapt to climate change, along with large corporations, small businesses, cities, households, and individuals. "Too small to make a difference" is neither scientifically accurate nor socially defensible.

Most importantly, climate change must be integrated into *all* conservation decisions. Soon after we began our service as director and science advisor, we found ourselves in Everglades National Park being briefed on a proposal to permit a hotel development near the Flamingo Visitor Center. The hotel would be constructed literally inches above sea level—in a location where sea level rise is sure to threaten coastal resources. It was not built, of course, but what is more important is that following this experience, climate change considerations were strategically built into all such development decisions for the National Park Service. Whether it is a county park planning board, historic preservation commission, tourism business, or large federal land management agency, climate change considerations must be formally and transparently integrated into decision-making processes.

Improve the training of future
conservation professionals

Training the next generation of conservation professionals is a paramount obligation of current conservation leaders. Contemporary training (often through traditional university degree programs) is compartmentalized, narrow in scope, and overly focused on operational or technical knowledge. This training must evolve and expand to include scientific skills, political advocacy tactics, and social justice ethics.

Scientific skills should include a sophisticated level of scientific literacy. This includes the capacity to apply scientific information to environmental and social problems, to manage science as a tool for litigation, and to maintain a healthy wariness to judge available research as to its quality, merit, and value. Such training would provide conservation leaders with first-hand experience in applying science, leading citizen science efforts, and communicating about science clearly and often with scientists, decision makers, and the public. Advocacy training regardless of party affiliation should include strategic planning, learning the social context in which the new professionals will work, operational skill in conducting political campaigns for causes and candidates, and inventive approaches such as use of social media to mobilize committed citizens.

Importantly, training the next generation of conservation leaders must include careful study and thoughtful practice of social justice ethics. The philosopher Peter Singer, in his famous essay *Famine, Affluence, and Morality*, reminds us that "acts are right or wrong by their consequences."[14] Future conservation leaders must be able to articulate *why* they are engaged in conservation activities and to what ends. For some, that will be a "deep ecology," focused on the consequences to nature. For others, it will be a focus on environmental justice, asking: Who benefits and who loses, and why? For still others, it will be intensely held religious beliefs about the responsible stewardship of the Creation. Regardless, the formal training and application of ethics is a critical part of the education and development of future conservation activists.

There are multiple delivery systems, each with its own strengths and weaknesses. American universities can provide much but not all of the needed training, but their subject matter silos, administrative bureaucracies, and aversion to risk make inventive training difficult. Affiliated institutes, with some independence yet able to tap faculty expertise, may be an effective solution. Intensive short-course experiences offered by universities, nongovernmental organizations, or even corporations can offer the benefit of highly focused training, the creation of cohorts of leaders that can support each other as they move through their careers, and the flexibility of field experiences. Individual organizations can create internal "academies," with committed mentors guiding young leaders in hands-on learning. Online courses can provide occasional technical training but are of limited value given the integrative demands of this educational mission.

There are other models: the ritual handing down of cultural traditions by Native American elders, the training of new musicians by old masters, apprenticing young persons to experienced craftspersons, writing workshops for beginning writers, and the mentoring of new academics by tenured professors. All of these models can be adapted for the creative training of conservation leaders.

Encourage and support conservationists to run for elected office at local, state, and national levels

In the American form of government, elected officials have substantial powers to influence the course of the nation; the laws and regulations we follow (or not) at federal, state, and local levels; and the priorities for spending taxpayer dollars. The importance of elections is evidenced by the 2016 election cycle. Elected climate change deniers can reject the overwhelming consensus of the scientific community and make policy choices that delay constructive responses and accelerate long-term harm. Other elected officials present the bogus "jobs versus the environment" argument as a binary choice and work to unwind years of conservation success, attacking the Endangered Species Act, the Land and Water Conservation Fund, historic preservation tax credits, and other legislation and initiatives.

While there is a relatively broad base of public support for conservation across the political spectrum, conservation values are not fully represented in the majority of local governments, state legislatures, governorships, and Congress.[15] The will of the people can and should be heard through protest and, if necessary, nonviolent civil disobedience, but it is most effectively expressed by those we elect. Conservation organizations have traditionally thrown support behind one candidate or another based on some expectation that when needed they will be supportive of conservation. These organizations have also worked to influence existing elected officials in the hope of advancing their conservation agendas. This influence-the-already-elected approach is increasingly falling on the deaf ears of elected officials, who question science, adhere to policies with no factual basis, and are beholden to extractive industries. What the conservation community has not adequately done is recruit new candidates who already have strong conservation values, help train them to run for office, and put collective support behind their campaigns. This should change.

There is a model of success that conservationists can emulate: EMILY's List. Ellen Malcolm, who founded EMILY's List in 1985, noted a significant dearth of women in elected roles. There were no women in the U.S. Senate, yet this governing body was regulating the lives and health of women across the nation. With the specific goal of electing pro-choice women to office, EMILY's List has become a model of grassroots work, recruiting and training women to run for office, raising funds for their support, and leading political strategies toward their ultimate election. Similar "Conservation Lists" should be developed, with a focus first on local elections, so that candidates gain the experience and credentials to run for higher elected offices.

To be elected in contemporary America, even at the local level, requires support, funding, political skill, media savvy, personal sacrifice, and a very thick skin. There is no reason that more individuals with strong conservation values, and/or professional training in science cannot gain these attributes and be successfully elected to office.

Execute an intergenerational transfer of power

The conservation movement is at a historical moment, a "tipping point" propelled by the aging of baby boomers, the maturing of Generation X, and the coming of age for millennials. In 2017, the baby boomers (born between 1946 and 1961) are between fifty-three and seventy years old.[16] The millennials are those born in the mid-1990s to early 2000s and are now in young adulthood. An intergenerational transfer of power is not only inevitable, but also urgent. We are not suggesting these young people be "listened to" or "at the table" or "constructively engaged," though all of those incremental tactics are worthy and valuable. We are arguing that the power to make decisions, lead organizations, advocate for causes, and direct the future of conservation belongs in new, diverse, and younger hands.

This transfer of power should be intentional, strategic, and occur at all levels of conservation. Young leaders (including those of the millennial generation) must step up and take on the challenge of directing the conservation movement. Older conservation leaders of the baby boom generation must be careful not to succumb to the "genetic method" of leadership turnover, selecting and promoting only those leaders that look, act, and lead as they have done. They must step out of the comfort of their traditional approaches to successional leadership and their conception of appropriate career paths and mentor promising young people who represent the new American demographic. Executive recruitment, job descriptions, hiring practices, workplace operations, and career ladders should be reformed to attract and select a new, diverse generation of leaders. Bravery is needed by all sides.

At the same time, there is a large reservoir of untapped collective wisdom in conservation organizations, built on decades of success and failure, conservation battles well fought, and passion for special places. The most successful have developed and tested strategies and skills fundamental to the fight: strategic intention, a clear set of objectives, access to high-quality science, a coherent political strategy, sophistication with the media, active public engagement, a strong legal defense, and sustained local presence. It is time that these skills and experiences are passed on, with the trust that this new generation will carry the conservation agenda forward with no less passion and with different and innovative approaches to the challenges that lie ahead.

*
**

These strategies, we believe, can help guide the future of conservation in America. Where they are already being employed, we applaud and encourage the conservation community to carry on. Where they are not yet attempted or applied only sporadically, we urge that they be considered, debated, adapted, and turned into specific tactics for conservation action.

Such strategies should not be considered disparate, isolated, or one-off approaches to achieving conservation goals. Each of the strategies can have a multiplier effect—applying one can amplify the effectiveness of others. For example, engaging the public in citizen science can increase support for environmental candidates running for public office, and improving the training of future conservation professionals will better prepare them to incorporate climate change into all dimensions of their decision making. In addition, it is the *collective* impact of these strategies that matters most. When they are used with strategic intention and combined for maximum effect, America can make significant conservation progress.

There is one more strategy to be suggested. On July 4,

1776, the Second Continental Congress meeting in Philadelphia adopted the Declaration of Independence, signed only by the Congress's president, John Hancock.[17] The eventual signers of this document knew well the dangers in declaring a new United States of America. Benjamin Franklin, it is said, remarked dryly: "We must, indeed, all hang together or, most assuredly, we shall all hang separately." Franklin's wisdom remains relevant to the American conservation movement. A new and unified vision of conservation is required.

CHAPTER FIVE

Toward a Unified Vision

Conservation in America currently resembles a deep-rooted tree at a mature stage of growth, with its older, large branches bent far apart. Nature conservation organizations and historic preservation groups share little in the way of strategies, tactics, or resources, and often (as in lawsuits by wilderness advocates to remove historic structures) actively oppose each other. The American health community is minimally involved in the protection of clean air and clean water, though continued protection of air and water is absolutely vital for American's health. Outdoor recreation and resource management agencies are only just now beginning to realize their potential as health-care providers capable of contributing to mental and physical well-being, aiding recovery from posttraumatic stress disorder, and reducing obesity. Environmental justice activists struggling to clean up urban neighborhoods are often disassociated from wildlife protection activists struggling to maintain biodiversity, missing opportunities to form rural-urban alliances as well as the potential of local communities to collectively serve as monitors and protectors of urban and regional environmental quality. The people of rural communities and regions often take their access to public lands for granted—and some support political agendas that would divest, develop, and degrade these resources even though there is little local economic benefit and their own access would be curtailed.

We argue that a more unified vision of what constitutes conservation is vital for the future of the conservation movement. This unified vision of conservation would include and integrate several major branches, each representing an "assemblage" (to use a term from ecosystem ecology) of citizens, activists, philanthropists, organizations, tribes, local, state, and federal agencies, business-sector (including the outdoor retail industry) and individual firms, scientists, and public leaders. We present these branches not in any priority order, as all are essential to the growth and progress of conservation.

The first branch is the traditional nature conservation community, dedicated to the preservation of biological diversity, in general, and individual species (often iconic ones like whales or wolves), in particular. These interests are often linked to conservation in specific and often highly contested locations—including sport hunting and fishing areas, tribal lands, wilderness, urban open space, marine reserves, and rare ecological systems such as the bioluminescent bays of Puerto Rico or the high alpine meadows of the Northern Rockies. Federal agencies such as the National Park Service and U.S. Fish and Wildlife Service play a central role, as do advocacy organizations such as the Sierra Club, the National Parks Conservation Association, and the League of Conservation Voters. Land trusts such as the Trust for Public Land and The Nature Conservancy, which buy, hold, and manage lands for conservation values are important contributors. Equally important stakeholders are tribal governments, especially of federally recognized tribes with established treaty rights. Nature conservation is often linked to the outdoor recreation, hunting, and fishing sectors dependent on these landscapes.

The second branch is the historical preservation community, which in the United States is largely focused on the preservation of historical structures such as the homes of famous Americans or buildings that bear witness to major events in American history, historic objects, and cultural landscapes such as battlefields or urban historical districts. Often, the historic preservation community is dominated by local or state interests, with national organizations such as the National Trust for Historic Preservation operating as facilitators, grant makers, and lobbyists with Congress. Both preservation science and contemporary scholarship (from history to cultural anthropology) can provide guidance to new interpretations of historical events. Historic preservation is often linked to the tourism industry by creating and managing "attractions" for local rural communities and metropolitan areas, in addition to preserving critically important historic resources, often supported by private philanthropy.

The third branch of a unified conservation includes local, state, regional, and national interests that advance the protection and provision of ecosystem services—the valued benefits society receives from functioning, healthy ecosystems. Examples include clean drinking water, medicinal resources, pest and flood control, carbon sequestration, and more. The community of interest often includes city, state, and federal planners working to maintain or improve all types of ecosystem services and regulators focused on prevention of harm, as well as scientists, tribal leaders, organic farmers, engineers, landscape architects, and powerful administrative organizations. An institutional example is the Metropolitan Water Reclamation District of Greater Chicago, an independent agency of state government with an elected board of commissioners and the responsibility of managing the water supply and wastewater treatment of the greater Chicago area. A direct-action example was the long-running protest against the Dakota Access Pipeline, led by Native Americans calling themselves Water Protectors, who had encampments near the Standing Rock Sioux Reservation and were bearing witness to the potential to harm the reservation's water supply.

The fourth branch embraces the environmental justice and civil rights communities. Environmental justice directly links access to a healthy living environment at the neighborhood or regional level to civil rights and protections. While often focused on identity politics and advancing the rights of individual groups (such as Black Lives Matter, tribal rights groups, or LGBTQ organizations), these interests also include organizations such as the Southern Poverty Law Center that work toward broader restorative justice, education, and resistance to hate groups (currently numbering over nine hundred) operating in the United States.[1]

The fifth branch of future-oriented conservation is the sustainability community, focused on developing sustainable infrastructure and technology, reducing unnecessary consumption, and creating sustainable livelihoods. Government and nonprofit organizations actively participate in this effort, often creating "green" development goals, plans, and small-scale enterprises, such as community gardens or recycling programs. The private sector (from the renewable energy industry to outdoor retail to building construction) plays a major role in developing and marketing sustainable options for the general American economy.

The sixth branch of conservation is the health community, which includes physicians, insurance companies, health providers, hospitals, and regulators charged with ensuring levels of quality and safety. The health community ranges from small town clinics with one general practice doctor to large corporations (such as Humana and UnitedHealth), to federal agencies such as the Food and Drug Administration and Centers for Disease Control and Prevention, to scientists and clinicians. Powerful professional associations such as the American Medical Association set standards, monitor misconduct, and recommend policies. The focus is on human health, often at the level of the individual patient and sometimes at the level of households or occupational groups. There is a smaller yet committed community of interest surrounding animal health— wildlife, domestic production animals, and the large U.S. pet population—and an emerging interest in One Health, a collaborative effort to "obtain optimal health for people, animals, and the environment."[2]

The seventh branch of conservation is the scientific community. As we have described throughout this book, science provides both information and insight essential to effective conservation. The scientific community is broad and varied in America and includes research universities, institutes for advanced studies, federal research programs, professional societies, and corporate laboratories. An example is the American Association for the Advancement of Science—the world's largest general scientific society, with over a hundred twenty thousand members and active in promoting the sciences across and among ninety-one partner organizations. Often these institutions are linked by agreements and shared activities. An example is the Cooperative Ecosystem Studies Network, which includes seventeen federal agencies and over three hundred fifty additional partners.[3] Nonprofit organizations like The Nature Conservancy or NatureServe maintain their own research staffs, and individual scientists are engaged in thousands of scientific projects that support conservation, from inventorying and monitoring small-site environmental conditions to large-scale ecosystem assessments.

*
**

All of these branches—nature conservation, historic preservation, protection of ecosystem services, environmental justice and civil rights, sustainability, health care, and science—are essential. Acting individually, each branch will be unlikely to achieve significant progress in the face of deliberate, determined, and direct attacks on conservation. United and working collaboratively, they are more capable of confronting the assault on conservation by the Trump administration, learning the strategic lessons of the current populist revival, and authentically and intentionally responding to the deep resentment, frustration, and fear among many Americans that the populist rising has exposed.

The challenges of conservation in the years ahead are too complex, wide-ranging, and consequential for one branch to take responsibility or "ownership." Division within the conservation community—to maintain outmoded traditions, argue over arcane terminology, protect "turf" or membership, further a single organization's goal or strategy, or prioritize independence over collective action—is not a viable option. What is necessary is an all-branches approach with intentional strategies for action, new alliances, and new forms of transformative change.

Several steps can be taken to advance this unified vision of conservation. Communication and sharing of agendas across the branches can be increased: meetings and conferences hosted by nonprofit organizations, government agencies, universities, and professional societies can strive to broaden participation beyond their traditional participants to include representatives from the other branches, as well as local interested citizens, and to foster new forms of collaboration. The national leaders of these branches of conservation should meet regularly, beginning with a national conservation summit, and work to establish shared strategies and coordinated action. As we have argued for an intergenerational transfer of power, young voices must be heard at these events.

The organizations that operate within each branch of conservation can expand their missions—beginning with examining and revising their formal mission statements and strategic plans. Key themes that can bind these organizations together include (but are not limited to) green urban renewal, environmental justice, and reduction of inequality. By broadening their "organizational directions" to legitimate and enable engagement across the branches of conservation, these groups can more easily support conservation's unified advance.

Collaborative conservation actions should be encouraged, particularly where distinctive advantages from working across branches are obvious. One example is climate change: all branches of conservation have a considerable stake in monitoring, mitigating, and adapting to the accelerated pace of climate change. Another is a broad and ambitious effort to treat conservation as a health issue, which can range from programs that link patients to parks to those protecting forests, oceans, and clear air and water as a public health response that benefits all Americans, not just the wealthy or privileged. A third example is access to well-protected public landscapes; witness the extraordinary decision by the outdoor retail industry to move their annual convention out of Utah because of the state's political opposition to the establishment of Bears Ears National Monument and its open hostility to the protection of public lands. Less obvious but urgently needed is collaboration among urban park managers, historic preservationists, and civil rights advocates to identify just solutions to the problem of Jim Crow–era Confederate monuments located in city public spaces—and not just in the American South but throughout the country as well.

The unified vision of conservation we call for reflects the challenging conditions of contemporary American society.

It encourages a convergence of interests at once necessary and powerful. It can create deep and meaningful common cause and calls out to the next generation of citizens. We believe that working toward this unified vision of conservation is perhaps the most important strategy of all to navigate the rough water ahead.

CHAPTER SIX

Resilience

The forces that propelled the populist election of Donald Trump pose grave threats to conservation in America. They challenge the reality and significance of climate change, which every major political, economic, environmental, and social decision in the decades ahead will need to consider. They rebuff the need for environmental protection, reject the scientific knowledge needed to make wise decisions, and rebuke those that would impose prudence and restraint on corporate use and misuse of precious common resources—public lands, clean air, clean water, thriving and sustainable cultural landscapes. The retrenchment and suppression of conservation at the federal level—without a strategic and coordinated resistance—may well set back conservation progress in the decades to come.

Yet as we have demonstrated, there are strategies and tactics that a determined and more unified conservation movement can apply, not simply to navigate these troubled times but also to make significant and sustained advances. We ground this belief in a firm faith in our national values, institutions, the conservation movement, and the American people. There are sound reasons for this long view.

There are deep American values that even now bind us together. Americans harbor a need to be respected as individuals and for their families to have futures equal to or better than the present can provide. They expect to have their lives matter, to not be forgotten (or abused) by their government, and to have a voice within their community and civil society. Americans live in vastly different locales—from urban to rural and tropical to arctic conditions. But most Americans share a deep sense of place—that the landscape they live in is an important part of their lives. Ranchers in West Texas, urbanites in West Seattle, and factory workers in West Virginia can all speak with fervent care (and sometimes love) for their portion of the American landscape.

A conservation movement that remembers these values and acts to embrace them is likely to be successful. Success cannot be measured as total agreement or lack of conflict. Americans *argue*; conservation decisions will always be highly contested because the outcomes keenly matter to us as citizens. Likewise, success cannot be measured as achieving a conservation goal (such as more wilderness, an improved management policy, or a new and sustainable product on the market) by overpowering the opposed, ignoring the minority, or avoiding the necessary "one thousand cups of coffee."

Conservation in the long view benefits from the resilience of American institutions at the local, state, and national level. From the New England township to the North Dakota county to Washington, DC, these institutions have over two centuries of experience during times of war, peace, hardship, and prosperity. The checks and balances designed by the Founders have become integrated into political life at all levels—local mayors checked by local city councils, state courts by federal district courts, the president by Congress, and all by the almost constant roil of elections at every level. Civil service protections for government workers, recall mechanisms for improvident (or unpopular) politicians, litigation in the courts, a free press that ranges from national news outlets to local weekly newspapers, and the cacophony of the internet all add to this national resilience.

These checks and balances mean that conservation progress is necessarily incremental rather than revolutionary. Successes are punctuated with disappointments; setbacks follow advances. We strive toward a "more perfect union," but we are not there yet. As Dorceta Taylor reminds us in *The Rise of the American Conservation Movement*, "The legacy of race and class discrimination and the practice of separating environmental issues from those of social inequality are challenges that the conservation movement has had a difficult time overcoming."[1] Conservation is not for the fainthearted or for those that seek guaranteed gratification or personal power. Compromise in support of first principles, combined with a sense of service and strategic intention, is a potent strategy that should be seen as honorable by the unified conservation movement. Following this strategy across the country, private citizens are offering their time and wealth to further conservation efforts, cities and towns are taking action to restore rivers and create new public spaces, and the private sector is stepping into leadership roles for conservation, sustainability, and climate response as federal government inaction creates a void.

We also have faith and optimism in the march of science, its advancing of human knowledge and insight, and the constant creation of new scientific tools to solve new problems. We acknowledge that, if misused, science can cause harm, but on balance its advancement will support the conservation movement. Our understanding of natural history, current social and environmental conditions, and human and ecological system functions is exponentially increasing. Regardless of how the current administration attempts to constrain climate science, such research will continue to be done and expand our understanding of climate change and the options for confronting this new and evolving planetary condition. The scientific community is finding its voice advocating for "evidence-based decisions," and individual scientists are becoming more skilled at communicating with the public.

We have confidence that the unified vision of conservation will result in significant progress over the long term. The coming together of nature conservation, historical preservation, ecosystem services, environmental justice and civil rights, sustainability, public health, and science communities is overdue, but when fully accomplished will reap significant reward. As these interests increasingly practice the skills of collaboration, and gain experience in working closely together in more common cause, they will find their collective "voice" to be powerful, influential, and effective. There will be a time when the physician, the pastor, the park ranger, the business leader, the scientist, and the school teacher all working together for conservation will seem not unusual but expected.

We have faith in and admiration for what the next generation of conservationists can and will accomplish. They inherit a world fraught with peril and a nation divided, but their inheritance also includes access to global knowledge, commitment to improving their communities, and self-awareness that they can and should empower themselves. Over our careers, we have met countless young persons interested in parks, history, education, politics, science, environmental justice, public service, or all of the above. Their enthusiasm for making a difference and impatience over waiting their turn is enlivening. They will assuredly make their own mistakes in actions for conservation, as we have, and will learn from them. If, as mentioned earlier, the future of conservation in America will depend on these exceptional young persons, our optimism for conservation is well-founded.

Finally, we are optimistic because we have seen firsthand the restorative powers of nature. If provided the opportunity, and sometimes assisted by human insight and skill, nature can recover and flourish. We have seen high meadows on the flanks of Mount Rainier return from bare ground to an eruption of alpine flowers. We have seen flows of water critical to what Marjory Stoneman Douglas called the "River of Grass" be replenished in the Everglades of Florida, and salmon once again swim the Elwha River from the sea to the mountains in the Pacific Northwest. We have seen community gardens bursting with vegetables in formerly vacant lots of Chicago and Baltimore, and the renewal of beaches and tropical forests on the island of Vieques, Puerto Rico, that was once strafed, bombed, cratered, and littered with munitions.

And it is not just nature that can recover and flourish. So can our history, our sense of justice, and our respect for civil rights. James Baldwin reminds us: "American history is longer, larger, more various, more beautiful, and more terrible than anything anyone has ever said about it."[2] We believe the story of America can be recast as more inclusive and truthful. We have seen the home of César Chávez take its place alongside the homes of American presidents. We have seen Harriet Tubman's Thompson Memorial African Methodist Episcopal Zion Church become part of the recognized national story. American history will, we believe, become more accurate, more inclusive, more diverse, more interesting, and more faithful to the deeper national sense of identity.

For all these reasons, we are confident in the future of conservation in America.

*
**

Throughout our careers we have witnessed many epic conservation actions. The battle to save California's Drakes Estero estuary from a commercial oyster farm in Point Reyes National Seashore ranks up there with the toughest. It is a story told with honesty and accuracy in Summer Brennan's *The Oyster War: The True Story of a Small Farm, Big Politics, and the Future of Wilderness in America*.[3] The oyster farm predated the establishment of the national seashore. The National Park Service recognized that commercial farming of nonnative oysters and clams (imported from Japan) and its associated infrastructure and motorboat use was incompatible with the natural functioning and official wilderness designation of the estuary.

In order to restore the estuary as a refuge for birds and harbor seals and its extraordinary sea grass beds, in 1972 the federal government purchased the Johnson's Oyster Company operation and its land-based facilities. As often is the case to allow the private sector to transition smoothly, the operator was granted a lease with a term of operation that expired in 2012, at which time the miles of oyster racks and processing buildings would be permanently removed. With only five years remaining on the permit, Johnson sold to new owners, the Drakes Bay Oyster Company. The new owners vowed to stay permanently.

Thus began a classic fight that pitted conservation and protection of wilderness against a damaging but historic use of the same resource. Members of the Senate and House of Representatives on different sides of the issue dueled, using their political power to influence the fate of the estuary. Political leadership in the Department of the Interior regularly called for the removal of the park superintendent who was standing up for the resource, but he was protected by his career supervisor. Local agricultural interests and park opponents, supported by a local newspaper that repeatedly attacked the National Park Service, filed formal complaints with the inspector general against agency leadership (including both of us). There were multiple allegations of scientific misconduct; all were determined to be untrue. In spite of the reputation of Marin County, California, as a bastion of environmental awareness, the issue deeply divided the community.

State jurisdiction over federal lands came into play, and lawyers on both sides debated the legal arguments. The oyster industry, supported by wealthy conservative-cause contributors, hired Washington, DC, lobbyists to initiate the politics of personal destruction against National Park Service leadership, scientists, and the environmental community. The legal case wound through the federal system all the way to the Supreme Court, which ultimately refused the case, remanding it back to the lower court, which called for a settlement by the parties. After an exhausting multiyear battle, the oyster farm closed in 2015, leaving behind over 3.8 million pounds of debris—treated lumber, discarded plastic, and old buildings—which the National Park Service agreed to clean up with the support of the National Park Foundation.

While every conservation struggle is unique, the commonality among the most difficult conflicts is that success involves dedicated organizations that thoughtfully and fearlessly advance conservation in ways that are intentional, intelligent, forceful, and relentless, incorporating many of the strategies we have described in this book. Yet individuals are often the true heroes of conservation. In the struggle over the future of Drakes Estero, Amy Trainer, a young and dedicated conservationist, was one of those individuals. Her persistent push, her quick but smart responses to every angle of attack, and her perseverance stand as a model for future conservationists. She navigated some very rough water with strategic intention, and Drakes Estero, the nation, and future generations are the beneficiaries.

In 2016, the year of the centennial and election, on one of the low sandbars of Drakes Estero, a harbor seal hauled out with her pup. For the first time in almost a hundred years, a harbor seal sunned undisturbed by motorboats run by oyster farmers. She does not know why it is so quiet, but we do.

ACKNOWLEDGEMENTS

We thank Christie Henry and Miranda Martin, our editors at the University of Chicago Press, for encouraging us to share our views on the immediate and long-term future of conservation in America. Maddie Duda provided valuable research and library support, and Alexis Ward prepared numerous drafts of the manuscript; their assistance is much appreciated. We are profoundly grateful to Terry Tempest Williams for taking precious writing time to prepare her moving foreword. Dave Harmon, Paula Jarvis, Shirley Malcom, Jaime Matyas, Gabrielle Names, and Ron Tipton reviewed and commented on early drafts; we thank them and several anonymous reviewers for their insights. Cynthia Barnes made numerous suggestions to bring concision and clarity to our writing, as did Yvonne Zipter. Of course, we alone are responsible for what resides between the covers.

NOTES

CHAPTER ONE

1. "Annual Visitation Summary, Report for 2016," National Park Service, accessed May 13, 2017, https://irma.nps.gov/Stats/SSRSReports/National%20Reports/Annual%20Visitation%20Summary%20Report%20(1979%20-%20Last%20Calendar%20Year); National Park Service, *Antiquities Act Designations and Related Actions: NPS Only.*

2. Donald Trump, Twitter Post, January 26, 2012, 1:40 P.M., https://twitter.com/realDonaldTrump.

3. David J. Voelker, introduction to Abraham Lincoln, "The Gettysburg Address," at http://historytools.davidjvoelker.com/sources/lincoln-gettysburg.pdf. See also Garry Willis, *Lincoln at Gettysburg: The Words That Remade America* (New York: Simon and Schuster, 1992).

4. Yoni Appelbaum, "Trump's Gettysburg Address," *Atlantic*, October 23, 2016, https://www.theatlantic.com/politics/archive/2016/10/trumps-gettysburg-address/505106/.

5. "Election 2016: Money Raised as of Dec. 31," *Washington Post*, February 1, 2017, https://www.washingtonpost.com/graphics/politics/2016-election/campaign-finance/.

6. See, for example, Dorceta E. Taylor, *The Rise of the American Conservation Movement: Power, Privilege, and Environmental Protection* (Durham, NC: Duke University Press, 2016).

7. See John B. Judis, *The Populist Explosion: How the Great Recession Transformed American and European Politics* (New York: Columbia Global Reports, 2016); Pankaj Mishra, *Age of Anger: A History of the Present* (New York: Farrar, Straus, and Giroux, 2017);

Jan-Werner Muller, *What Is Populism?* (Philadelphia: University of Pennsylvania Press, 2016); and Donald I. Warren, *The Radical Center: Middle Americans and the Politics of Alienation* (Notre Dame: University of Notre Dame Press, 1976).

CHAPTER TWO

1. Horace M. Albright and Marian Albright Schenck, "The Mather Mountain Party, 1915," in *Creating the National Park Service: The Missing Years* (Norman: University of Oklahoma Press, 1999), https://www.nps.gov/parkhistory/online_books/albright2/pdf/ch7.pdf.

2. "Organic Act of 1916" (Title 16, USC 1.1A), National Park Service, last updated February 5, 2017, https://www.nps.gov/grba/learn/management/organic-act-of-1916.htm.

3. Prior to the 1960s, only a few wealthy individuals like John D. Rockefeller had donated funds to the National Park System. Rockefeller, for example, donated land for Virgin Islands, Acadia, Grand Teton, and Great Smokey Mountains National Parks.

4. Richard Nixon, "Special Message to the Congress Outlining the 1972 Environmental Program," February 8, 1972, cited online by Gerhard Peters and John T. Woolley, The American Presidency Project, http://www.presidency.ucsb.edu/ws/?pid=3731.

5. Daniel Nelson, *Nature's Burdens: Conservation and American Politics, the Reagan Era to the Present* (Boulder: University Press of Colorado, 2017).

6. National Park Service, *A Call to Action: Preparing for a Second Century of Stewardship and Engagement* (Washington, DC: National Park Service, U.S. Department of the Interior, 2011), https://www.nps.gov/calltoaction/PDF/C2A_2014.pdf.

7. Richard Louv, *Last Child in the Woods* (New York: Workman Publishing, 2005).

8. Children's Defense Fund, "Child Poverty in America 2015:

National Analysis," September 13, 2016, http://www.childrensdefense.org/library/data/child-poverty-in-america-2015.pdf.

9. For more on the 21st Century Conservation Service Corps, see www.21csc.org.

10. By the end of 2016, the campaign exceeded sixteen billion media impressions and stimulated over three thousand events, books, videos, art projects, media articles, and park visits. National Park Service, *Realizing the Vision For the Second Century: National Park Service Centennial Final Report*, December 2016, https://www.nps.gov/subjects/centennial/upload/Centennial-Final-Report-December-2016-single-pages.pdf.

11. See, for example, E. O. Wilson, *Naturalist* (Washington, DC: Island Press, 1994); or Douglas Brinkley's biographies of Theodore Roosevelt (*The Wilderness Warrior: Theodore Roosevelt and the Crusade for America* [New York: Harper Perennial, 2010]) or Franklin D. Roosevelt (*Rightful Heritage: Franklin D. Roosevelt and the Land of America* [New York: Harper, 2016]), and his essay "Thoreau and the Legacy of Wilderness," *New York Times*, July 9, 2017.

12. *Revisiting Leopold: Resource Stewardship in the National Parks*, National Park Service Advisory Board Report 2012, August 25, 2012, https://www.nps.gov/calltoaction/pdf/leopoldreport_2012.pdf.

13. Gary Machlis and Marcia McNutt, "Scenario-Building for the Deepwater Horizon Oil Spill," *Science* 329, no. 5995 (2010): 1018–19, DOI: 10.1126/science.1195382.

CHAPTER THREE

1. Waldo Frank, *Chart for Rough Water; Our Role in a New World* (New York: Doubleday, 1940), and *The Re-discovery of America: Chart for Rough Water* (New York: Duell, Sloan and Pearce, 1947).

2. Thomas Stocker et al., eds., *Climate Change 2013: The Physical Science Basis, Contribution of Working Group 1 to the Fifth Assessment Report of the Intergovernmental Panel on Climate Change* (Cambridge:

Cambridge University Press, 2014); Steven J. Smith, James Edmonds, Corinne A. Hartin, Anupriya Mundra, and Katherine Calvin. "Near-Term Acceleration in the Rate of Temperature Change," *Nature Climate Change* 5 (2015): 333–36, DOI: 10.1038/nclimate2552; Kevin Schaefer, Tingjun Zhang, Lori Bruhwiler, and Andrew P. Barnett, "Amount and Timing of Permafrost Carbon Release in Response to Climate Warming," *Tellus* B 63, no. 2 (2011): 165–80.

3. Union of Concerned Scientists, *When Rising Seas Hit Home*, July 2017, www.ucsusa.org/RisingSeasHitHome; J. N. Gregory, *The Southern Diaspora: How the Great Migrations of Black and White Southerners Transformed America* (Chapel Hill: University of North Carolina Press, 2005).

4. John Walsh, Donald Wuebbles, Katherine Hayhoe, James Kossin, Kenneth Kunkel, Graeme Stephens, Peter Thorne, et al., "Our Changing Climate," in *Climate Change Impacts in the United States: The Third National Climate Assessment* (Washington DC: U.S. Global Change Research Program, 2014).

5. N. Fann, T. Brennan, P. Dolwick, J. L. Gamble, V. Ilacqua, L. Kolb, C.G. Nolte, T. L. Spero, and L. Ziska, "Air Quality Impacts," in *The Impacts of Climate Change on Human Health in the United States: A Scientific Assessment* (Washington, DC: U.S. Global Change Research Program), , 69–98, http://dx.doi.org/10.7930/J0GQ6VP6.

6. Elizabeth Kolbert, *The Sixth Extinction: An Unnatural History* (London: Picador, 2015).

7. Gerardo Ceballos, Paul R. Ehrlich, and Rodolfo Dirzo, "Biological Annihilation via the Ongoing Sixth Mass Extinction Signaled by Vertebrate Population Losses and Declines," *Proceedings of the National Academy of Sciences* 114, no. 30 (May 2017): E6089–E6096.

8. Tracie McMillian, K. Cahana, S. Sinclair, and A. Toensing, "The New Face of Hunger," *National Geographic* 226 (2014): 66–89.

9. Solomon Hsiang, Robert Kopp, Amir Jina, James Rising, Michael Delgado, Shashank Mohan, D. J. Rasmussen, Robert Muir-Wood, Paul Wilson, Michael Oppenheimer, Kate Larsen, and Trevor

Houser, "Estimating Economic Damage from Climate Change in the United States," *Science* 356, no. 6345 (2017): 1362–69.

10. "Income Inequality in the United States," Inequality.org, last accessed May 11, 2017, https://inequality.org/facts/income-inequality/.

11. Thomas Shaprio, *Toxic Inequality: How America's Wealth Gap Destroys Mobility, Deepens the Racial Divide, and Threatens Our Future* (New York: Basic Books, 2017).

12. Noam Chomsky, *Requiem for the American Dream: The 10 Principles of Concentration of Wealth & Power* (New York: Seven Stories Press, 2017).

13. Amanda Erickson, "Trump's Climate Change Shift Is Really about Killing the International Order," *Washington Post*, March 29, 2017, https://www.washingtonpost.com/news/worldviews/wp/2017/03/29/trumps-climate-change-shift-is-really-about-killing-the-international-order/?utm_term=.2502299c602b.

14. Tom Nichols, "How America Lost Faith in Expertise: And Why That's a Giant Problem" *Foreign Affairs* (March–April 2017).

15. Stephen Harper was Canada's prime minister from 2006 to 2015; see Chris Turner, *The War on Science: Muzzled Scientists and Willful Blindness in Stephen Harper's Canada* (Vancouver: Greystone Books, 2014).

16. Kendrick Frazier, *Science under Siege: Defending Science, Exposing Pseudoscience* (Amherst, NY: Prometheus Books, 2009); Chris Mooney, *The Republican War on Science* (New York: Basic Books, 2006); Shawn Otto, *The War on Science: Who's Waging It, Why IT Matters, What We Can Do about It* (Minneapolis, MN: Milkweed Editions, 2016).

CHAPTER FOUR

1. Tony Blair, "Against Populism, the Center Must Hold," *New York Times*, March 3, 2017.

2. Cary Funk and Lee Rainie, "Public and Scientists View on Science and Society," Pew Research Center, January 29, 2015.

3. Arlie Russell Hochschild, *Strangers in Their Own Land: Anger and Mourning on the American Right* (New York: New Press, 2016).

4. Pope Francis, *Encyclical on Climate Change and Inequality: On Care for Our Common Home* (New York: Melville House Publishing, 2015).

5. Rebecca Skloot, *The Immortal Life of Henrietta Lacks* (New York: Broadway Paperbacks, 2011); Margot Lee Shetterly, *Hidden Figures: The American Dream and the Untold Story of the Black Women Mathematicians Who Helped Win the Space Race* (New York: William Morrow, 2016).

6. Russell Thornton, *American Indian Holocaust and Survival: A Population History since 1492* (Norman: University of Oklahoma Press, 1990).

7. *The National Parks and Public Health: A NPS Healthy Parks, Healthy People Science Plan*, National Park Service (Washington, DC: U.S. Department of the Interior, July 2013), https://www.nps.gov/public_health/hp/hphp/press/HPHP_Science%20Plan_accessible%20version.final.23.july.2013.pdf; Florence Williams, *The Nature Fix: Why Nature Makes Us Happier, Healthier, and More Creative* (New York: W. W. Norton, 2017).

8. "World Database on Protected Areas," United Nations Environment Programme: World Conservation Monitoring Centre, accessed May 14, 2017, http://old.unep-wcmc.org/world-database-on-protected-areas_164.html; "Marine Protected Areas," Our Ocean 2016, accessed June 12, 2017, http://ourocean2016.org/marine-protected-areas/.

9. E. O. Wilson, *Half-Earth: Our Planet's Fight for Life* (New York: Liveright, 2017).

10. The Eden Place Nature Center is an award-winning program run by the Fuller Park Community Development Corporation, a 501(c)3 nonprofit organization (http://www.edenplacenaturecenter.org).

11. J. Morgan Grove, Mary Cadenasso, Steward T. Pickett, Gary E. Machlis, William R. Burch Jr., and Laura A. Ogden, *The Baltimore School of Urban Ecology: Space, Scale, and Time for the Study of Cities* (New Haven, CT: Yale University Press, 2015). For more examples of the effects of green infrastructure, see "Crime and Green Infrastructure," Baltimore Ecosystem Study, https://www.beslter.org/frame4-page_3d_33.html; Julian Spector, "Another Reason to Love Urban Green Space: It Fights Crime," *Citylab*, April 13, 2016, http://www.citylab.com/cityfixer/2016/04/vacant-lots-green-space-crime-research-statistics/476040/; Christopher Coutts and Micah Hahn, "Green Infrastructure, Ecosystem Services, and Human Health," *International Journal of Environmental Research and Public Health* 12, no. 8 (2015): 9768–98; doi: 10.3390/ijerph120809; and Christopher Coutts, *Green Infrastructure and Public Health* (London: Routledge, 2016).

12. Timothy Snyder, *On Tyranny: Twenty Lessons from the Twentieth Century* (n.p.: Tim Duggan Books, 2017).

13. Charles Lindblom, *Usable Knowledge: Social Science and Social Problem Solving* (New Haven, CT: Yale University Press, 1979); *Revisiting Leopold*.

14. Peter Singer, *Famine, Affluence, and Morality* (Oxford: Oxford University Press, 2015).

15. For example, the Colorado College has conducted public opinion polls in the Western states (many conservative) of Arizona, Colorado, Montana, Nevada, New Mexico, Utah, and Wyoming. In 2017, the poll showed 80 percent support for keeping recent national monument designations in place, consistent across partisan, ideological, and state lines. See "The 2017 Colorado in the West Poll," Colorado College State of the Rockies Project, January 2017, https://www.coloradocollege.edu/other/stateoftherockies/conservationinthewest/index.html; and Shorna R. Broussard, Camille Washington-Ottombre, and Brian K. Miller, "Attitudes toward Policies to Protect Open Space: A Comparative Study of

Government Planning Officials and the General Public," *Landscape and Urban Planning* 86, no. 1 (2008): 14–24.

16. D'Vera Cohn, "A Demographic Portrait of the Millennial Generation," Pew Research Center, February 24, 2010, http://www.pewsocialtrends.org/2010/02/24/a-demographic-portrait-of-the-millennial-generation/.

17. Pauline Maier, *American Scripture: Making the Declaration of Independence* (New York: Vintage Books, 1997).

CHAPTER FIVE

1. "Hate Groups" (active hate groups in the United States), Southern Poverty Law Center 2015, accessed May 15, 2017, https://www.splcenter.org/hate-map.

2. "One Health Basics," Centers for Disease Control and Prevention, last updated August 4, 2017, https://www.cdc.gov/onehealth/basics/index.html.

3. "About AAAS," American Association for the Advancement of Science, last updated April 7, 2017, https://www.aaas.org/about/mission-and-history; "Network Resources," Cooperative Ecosystem Studies Units National Network, accessed May 15, 2017, http://www.cesu.psu.edu/materials/default.htm.

CHAPTER SIX

1. Dorceta E. Taylor, *The Rise of the American Conservation Movement: Power, Privilege, and Environmental Protection* (Durham, NC: Duke University Press, 2016).

2. James Baldwin, "A Talk to Teachers," *Saturday Review*, December 21, 1963, 42–44.

3. Summer Brennan, *The Oyster War: The True Story of a Small Town, Big Politics, and the Future of Wilderness in America* (Berkeley, CA: Counterpoint Press, 2015).

PHOTO CAPTIONS

PAGE 2: Crowd Arriving at Dedication of Gettysburg Battlefield, November 19, 1863. Courtesy of the Library of Congress.

PAGE 12: Spruce Tree House, Mesa Verde National Park. Photo credit: QT Luong/terragalleria.com.

PAGE 30: Civil Rights March on Washington, DC, August 28, 1963. Courtesy of the National Archives.

PAGE 42: Bison, Grand Teton National Park. Photo credit: Capture Light/shutterstock.com.

PAGE 68: Great Smoky Mountains National Park. Photo credit: Beth Ruggiero-York/shutterstock.com.

PAGE 78: Parade, Breckenridge, Colorado, July 4, 2014. Photo credit: Jessie Unruh.